AF863831

LIVERPOOL HOPE
UNIVERSITY COLLEGE
LIBRARY
HOPE PARK
LIVERPOOL, L16 9JD

Seed and **Surplus**

An illustrated guide to the **world food system**

Bertrand Delpeuch

translated by
Chantal Finney and **Alistair Smith**

CIIR / Farmers' Link

LIVERPOOL HOPE
UNIVERSITY COLLEGE
Order No./Invoice No. £8.14
L7281 / B224674
Accession No.
147951
Class No.
911.32 DEL
Control No.
ISBN
Catal.
7/97

First published in English 1994

Catholic Institute for International Relations (CIIR)
Unit 3, Canonbury Yard
190a New North Road
London N1 7BJ, UK

Farmers' Link
38-40 Exchange Street
Norwich NR2 1AX, UK

English Translation © Catholic Institute for International Relations (CIIR) 1994
All rights reserved

British Library Cataloguing-in-Publication Data.
A catalogue record for this book is available from the British Library.

ISBN 1 85287 125 3

Originally published in French as *L'enjeu alimentaire Nord-Sud*
© Syros-Alternatives, Paris, 1992. Also published in Spanish, Italian, Japanese and Portuguese.

Cover photograph © Jofre Masceno
Cover and text design by Jan Brown Designs
Printed in the UK by Russell Press Ltd

Contents

Foreword vii

Preface 1

Introduction 4

Acronyms and statistics 6

Chapter One

Finding out the facts 7

Malnutrition: A daily reality 8

Producing food is not the same as feeding people: The Brazilian paradox 10

Some Third World countries have become **major agricultural importers** 12

Africa's growing **food dependency** 14

A few developed countries account for the bulk of **world food trade** 16

The European Community: the world's biggest importer of agricultural products 18

The EC imports twice as much as it exports to the Third World in agricultural goods 20

The United States plays a leading role in world agricultural trade 22

Japan, with a deficit in animal feeds, is the United States' biggest agricultural consumer 24

Agricultural production is increasingly costly 26

The food industry processes a growing proportion of agricultural products 28

Around 20 companies account for the bulk of trade in agricultural products 30

US companies dominate the world food industry 32

A few multinational companies now control the **seed industry** 34

World market prices of agricultural commodities are unstable 36

Debt repayments undermine the growth of developing countries 38

Overseas development aid remains below international targets 40

The percentage of farmers is falling but not their overall number 42

Women play a crucial role in food production in the Third World 44

Millions of rural people have no land to work 46

The Western diet is spreading in the Third World from urban centres 48

The disappearance of tropical forests is threatening the eco-system 50

Over-intensive agriculture often neglects the environment 52

Chapter Two

Understanding the facts 55

The developed countries have replaced the Third World as major agricultural exporters 56

Food interdependence between countries is increasing rapidly 58

Decision-making is further and further removed from farmers 60

Governments play a crucial role in the organisation of world agricultural trade 62

'Green oil' seeks solvent markets 64

International commodity agreements face an impasse 66

New technologies increase competition between agricultural raw materials 68

Multinational companies profit from the differences between countries 70

The agricultural systems of developing countries are penalised 72

African governments have long discouraged their farmers by favouring imports 74

Foreign debt repayments reduce the food security of the poorest 76

Developing countries' food imports tend to increase with urbanisation 78

Cash crops often use productive resources at the expense of subsistence crops 80

Food aid often discourages local production 82

The use of cereals for animal feed is gradually spreading in the developing countries 84

Industrial fishing is costly in capital and energy and is overexploiting coastal resources 86

Biological progress: Who will benefit? 88

Advertising alters consumer behaviour without concern for local priorities 90

Environmental degradation often has social causes 92

Forced collectivisation has failed in agriculture 94

Chapter Three

Acting on the Facts 97

Give **Third World farmers** a chance 98

Carry out **land reform** to boost small farmer agriculture 100

Diversity agricultural development models 102

Promote **appropriate production systems** 104

Take **the role of women** into account 106

Opt for **triangular food aid** 108

Relieve the burden of **external debt** 110

Encourage **international codes of conduct** 112

Promote the **participation of social actors** in the South 114

Conclusion 117

References 118

Foreword

An estimated 35,000 people die of hunger every day. Yet over the last seven years the United States and the European Union have argued over how, and by how much, they should reduce their vast surpluses of agricultural products. Their dispute delayed the conclusion of the Gatt world trade talks until the end of 1993.

For many people, media reports of hunger and surpluses are confusing, and the connection between the two is rarely made. If some countries in the North of the world are producing more food than they can eat, why are people starving elsewhere? And why are agricultural systems in the Third World failing?

Explaining the causes of world hunger, drawing out the links with Northern agricultural and rural development policies, and promoting public debate in the North about the sustainability of our own food system are key aims of the work of both CIIR and Farmers' Link. In an effort to explain the issues restricting the access of millions of people across the world to such a crucial item as food, we have collaborated in the publication of **Seed and Surplus: An Illustrated Guide to the World Food System**.

A book on the world food system is, regrettably, as necessary and topical today as it was when the first edition of this updated volume was published in French in 1988. Although there have been 'success stories' in the struggle to end hunger, where large numbers of people have secured their own food supply, the threat of large-scale food crises still hangs over the 1990s and beyond. Despite past commitments by the international community, the 21st century is likely to dawn with around one billion people hungry.

The many injustices of the so-called 'world food system' – which has systematically failed to feed the world – are a prime cause of continuing hunger. And in many countries, hunger is all too often the consequence of small farmers being economically exploited or denied land rather than a problem of food scarcity. This book demonstrates, in simple but not simplistic terms, the unpalatable truths about the production, distribution and control of food.

The European Union, for example, has long been criticised for subsidising exports under its Common Agricultural Policy (CAP). As well as depriving developing countries of foreign exchange by reducing their market shares and depressing world prices, subsidised EU exports have undermined food systems in the Third World by flooding local markets with cheap imports. This has discouraged investment in agriculture.

Under last December's Gatt agreement, Gatt members committed themselves to increasing access to their markets for agricultural commodities, and to reducing export subsidies and domestic support for agriculture. In theory, this should raise world market prices of the major traded agricultural commodities, as lower subsidies and smaller surpluses reduce excess supply in world markets. In turn, higher prices would benefit competing agricultural exporters. For developing countries that import food, higher prices would make an increase in food aid and support for agricultural development vital in the medium term; but they could provide an incentive for governments to implement policies to increase the amount of food grown and sold by poor people, which is the key to improving the food security of the poor in many of the world's poorest countries.

At the same time, however, the Gatt agreement has been framed to allow both the EU and the United States to maintain support for their producers. Direct payments to farmers, as used in the US arable sector and the EU arable and livestock sectors, will be excluded from commitments to reduce domestic agricultural support. And critically, the Gatt deal will not stop export dumping: in effect, it will sanction EU subsidies on exports equivalent to almost 80 per cent of the massive surplus capacity represented by production in 1986-90.

In Europe, it is again vulnerable small farmers and rural workers who have most reason to fear from a 'shake-out' of EU farming: over 1,000 farming families have left the land every day for the last 30 years.

In recent years, however, there has been growing concern in Europe about the way food is produced. For while the quantity of food produced has gone up, the quality has gone down. Along with many consumers, thousands of farmers across the continent now advocate a different kind of agriculture, one that is less reliant on external inputs and conserves resources. It is their belief that such an agriculture, by reducing output, would also reduce the damage caused to Third World farmers by the subsidised exports of large-scale industrial agriculture. The CAP was designed to end hunger in Western Europe after the Second World War, but its sell-by date has long since passed. The time has come for an alternative agriculture that is both environmentally sustainable and internationally just.

In 1992, two historic inter-governmental conferences gave a new sense of urgency and commitment to the challenge of feeding humankind both adequately and sustainably. The UN Conference on Environment and Development's 'Agenda 21', signed by 178 governments in Rio de Janeiro, followed by the International Conference on Nutrition in Rome, both drew up action plans for tackling food imbalances and unsustainable agriculture. Unfortunately, both conferences fudged the issue of whether the plans were to be applied by the rich industrialised countries, as well as by Third World countries.

With the EU, the United States and Japan due to continue reform of their own much-criticised farm policies, and with the need for an alternative agriculture now appearing on the international agenda, the messages of **Seed and Surplus** will remain topical for some years to come. In common with the author, CIIR and Farmers' Link hope that the reflection and action provoked by this book will make an important contribution to the challenge of feeding the world.

CIIR / Farmers' Link
May 1994

Preface

LIVERPOOL HOPE UNIVERSITY COLLEGE

The world can feed itself – but it does not. It should be able to do so – but how? That is the subject of this book.

It would be easy to demonstrate that the world can feed all of its inhabitants and support a population of between 7 billion and 8 billion people without damage to the planet. There is enough land, water and sun, and there are enough farmers, tools and techniques for everyone to eat their fill. The scope for further progress is still considerable – in the developed countries, but even more clearly so in the countries of the Third World. It could even be said, without fear of contradiction, that world agriculture already produces enough to satisfy all hunger in the world, and that whatever food is lacking is in fact destroyed for want of proper storage or transport.

Yet hunger exists and is becoming more widespread. Food security is rare. This should not be seen as a global concept, but as a regional, national, even local one. It is of little help to someone dying of hunger in Calcutta to know that the silos in Chicago are overflowing with an abundant harvest. It is of little help to Burkina Faso if the countries of the North are burdened with huge surpluses. A poverty-stricken person living on the banks of the Ganges does not live off statistics, but off a chapatti, a bowl of rice, a dried fish or a vegetable rich in vitamins.

It would be no more difficult for each country to produce enough for itself, than for the big powers of the Northern hemisphere to be able to feed the whole world from their harvests. Each country would not endure the blackmail of hunger, a blackmail which has often been envisaged by some.

Although countries like Egypt and Japan can clearly never be self-sufficient – their populations are too big in relation to the size of their land – the countries which could produce enough to satisfy their own needs far outnumber those which could not. So why are they unable to do so?

The reasons are mostly political.

To a large extent, the trade-off in many countries between urban and rural populations is a cause of food imbalances. Governments, in order to satisfy urban workers, civil servants, soldiers and the unemployed who live in the suburbs of capital cities, impose unremunerative prices on farmers. So farmers stop producing more than they need for their own subsistence, and young people leave for the cities to escape the sad fate of a countryside without prospects.

It is also so easy and handy not to have to go and collect harvests from the field when food aid and cheap cereals arrive at ports right nearby . Eating habits quickly take hold which make urban consumption dependent on imported foods that the country's land and climate cannot produce.

Thus, out of short-term concern, the agriculture of some countries is gradually destroyed. But this is done with the complicity of the 'generous' donors of the North and the active support of the market system. For the delivery price of US or European agricultural surpluses at an African port is usually significantly lower than the lowest cost price of these surpluses: the aid received by farmers in the North deprives farmers in the South of any chance of development.

Then there is doctrine. Liberal economics does not stop aid from being given to white farmers but does stop black farmers from protecting themselves against this unfair competition. Free trade, thus conceived and practiced, pushes the countries of the South into a food deficit which is growing by the day.

Though the reasons for the present situation are clear, this does not mean that the solutions are easy to implement. Indeed, the development of a modern agriculture in the South that can feed people involves a number of demands.

- The free trade creed must be open to challenge when it threatens food security. The South will only be able to develop its agriculture under the shelter of reasonable but effective protection.
- Southern governments must protect their domestic food production, and develop their rural sector, emphasising support for farmers.
- Northern governments must stop using their own agriculture as a strategic economic weapon.
- Aid for farmers in the North – which is necessary – must be adapted to the situation of food self-sufficiency now enjoyed by their countries.
- Food and agricultural research must be developed and diversified to address the needs of the South better, not just because of its soil and cli-

mate but also for more fundamental reasons: Northern agriculture consumes 10 energy calories for every food calorie produced. This model cannot prevail on a global scale. The earth and its subsoil cannot produce enough energy to sustain it in the long term.

- Finally, the role of agriculture must be revalued. The farmer is not just a food producer; she or he also conserves the natural environment and is the guardian of ecological balances which are today under threat.

This is why the programmed death of agricultural systems in the South should not be a source of satisfaction: it not only threatens peace, but the very fate of our planet.

Edgard Pisani

Edgard Pisani, a former adviser to French President François Mitterrand, was the EC's agriculture commissioner when the Common Agricultural Policy was created in the 1960s.

Introduction

Since the first edition of this book in French, in 1989, the world has seen considerable changes, including the collapse of the Soviet empire, a fragile process of democratisation in Africa and the Gulf War. But the reality of food inequalities remains.

In spring 1992, the European Community (EC) adopted a thoroughgoing reform of its Common Agricultural Policy (CAP). Only time will tell whether the proclaimed objectives of controlling production and protecting the environment will be met. Meanwhile, the Community seems very prepared to continue selling on the world market cereals which it cannot sell within its own borders.

Like an unbearable flashback, images of refugees driven to hunger by armed conflicts continue to reach us from the Horn of Africa, Kurdistan, Mozambique... Malnutrition, on the other hand, is less spectacular and continues to deprive millions of humans of their dignity.

Let us act, let us react. No fatalism is acceptable when these are the stakes. Yet while emotion must keep us from resignation, it can also be ill-advised. World hunger cannot be used to justify food surpluses in the North, when these are achieved at the cost of massive imports of cattle feed, of polluting chemical inputs and of deterioration of the natural environment. Western surpluses for empty African stomachs – if only it were so simple and convenient! But the reality is more complex. The purpose of this book is to present that reality in terms which are simple yet not simplistic.

The chosen format is inspired by that adopted by Rudolf Strahm in his book *Pourquoi sont-ils si pauvres?* ('Why are they so poor?').[1] Each left-hand page is a diagram. The facing right-hand page explains the idea and gives examples. The book is divided into three parts:

- Chapter 1 presents the basic facts about agriculture, food and trade in agricultural products worldwide;
- Chapter 2 seeks to explain the major trends at play in the North-South food trade;
- Chapter 3 makes proposals for action in both North and South.

To make for easier reading, examples appear at the bottom of the page and references at the end of the book.

Should the term 'Third World' have been used in such a book? Today, some writers question this expression on the grounds that it encompasses too great a diversity of countries. Others argue that it should be retained because it recalls colonial history and refers us back to two characteristics which are common to these countries, namely high population growth and major social inequalities.[1] Because it has become established through usage, the term 'Third World' has been retained in many of the diagrams. But wherever possible, a distinction has been made between the main geographical and economic blocs.

Such a book cannot claim to be complete; nor can it go into all the details which an analysis of social relations would entail. The priority given to the diagrams meant that choices were necessary; the scope of the subject matter meant that each commentary had to emphasise the essential facts. This handbook is above all an invitation to further reading, given that the cold statement of facts can never replace the testimonies of people like Domitila, Rigoberta Menchú, Oumar Dia or Robert Linhart.[2]

It remains for me to thank all those who have helped with their advice in writing this book, in particular the members of Solagral,[3] Alain Le Sann, Joseph Rocher, and René Schoenhenzel who drew the diagrams.

ACRONYMS

ACP	African, Caribbean and Pacific group of countries
CAP	Common Agricultural Policy
DAC	Development Assistance Committee
EAGGF	European Agricultural Guarantee and Guidance Fund
EC	European Community
ECU	European Currency Unit (Ecu1.00 = UK£0.80 on 30 April 1993)
FAO	Food and Agriculture Organisation
GATT	General Agreement on Tariffs and Trade
GDP	Gross Domestic Product
GNP	Gross National Product
IMF	International Monetary Fund
LDC	Least developed country
NGO	Non-governmental organisation
NIC	Newly-industrialised country
OECD	Organisation for Economic Cooperation and Development
OPEC	Organisation of Petroleum Exporting Countries
STABEX	System of Stabilisation of Export Earnings
UNCTAD	United Nations Conference on Trade and Development
UNHCR	United Nations High Commission for Refugees
UNICEF	United Nations Children's Fund
USDA	United States Department of Agriculture
WHO	World Health Organisation

STATISTICS

Wherever possible, the latest available figures have been used in this new English-language edition. There are however subjects (food availability, peasant landlessness etc.) where the most recent international surveys are a few years old. Furthermore, the figures for Third World countries should be seen not so much as statistical certainties, but rather as indicators.

The existence of large-scale illegal trade between African countries,[1] and the sheer scale of the non-cash economy in developing countries, mean that caution should be exercised when interpreting national statistics. Caution is also required when looking at comparisons between continents which are based on a single criterion. For example, the 'apparent' food dependency of India, measured in terms of cereal imports per capita, dropped from 9kg to 0.3kg between 1960 and 1980. In Africa, this same dependency has increased over the same period from 4.5 to 13.4 kg.[2] But this concept of 'apparent' dependency puts India in a false light in relation to Africa, obscuring as it does the fact that India's more intensive agriculture requires imports of agricultural inputs (fertilisers, pesticides, seeds) and know-how.

1. FINDING OUT THE FACTS

The diversity of food situations in the world

The agricultural trade of the major industrialised countries

The role of the agri-food industry

Who works the land?

Neglect of the environment

Malnutrition: a daily reality

Per capita food availability in daily calories

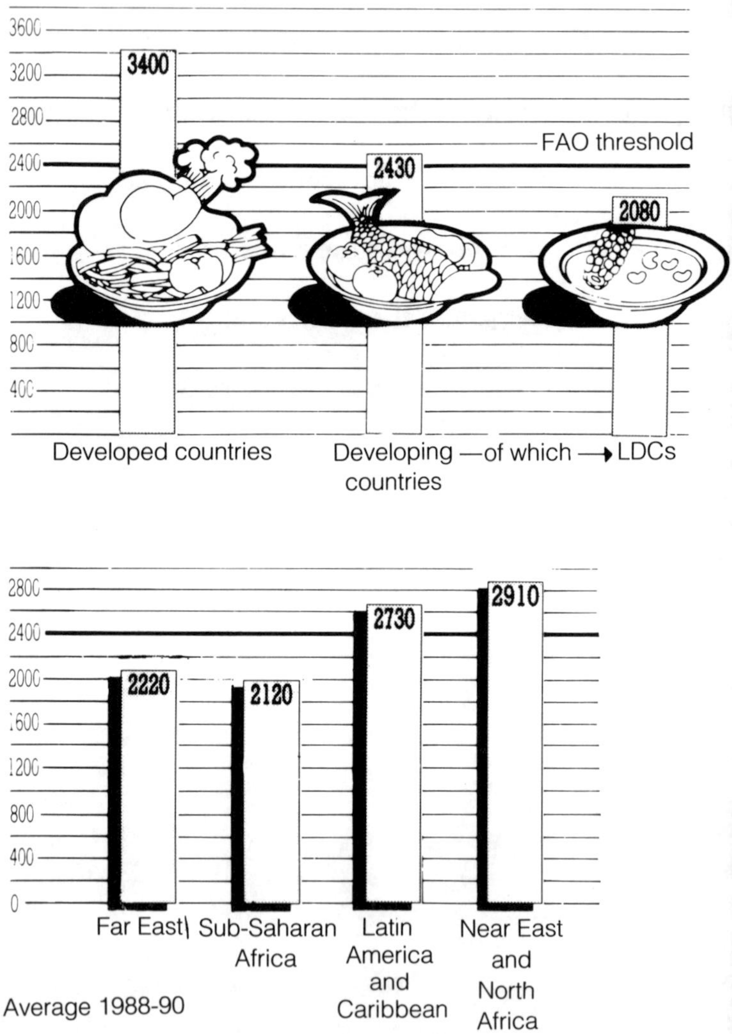

The FAO estimates an adequate daily energy intake to be 2,400-2,500 calories.

© Delpeuch – Solagral

Source: FAO

Worldwide, more than 700 million people suffer acute undernourishment; their total number is unlikely to change much before the end of the century. It is in South Asia that there is the greatest concentration of people suffering from hunger, and it is in Africa that their situation is deteriorating fastest. Some 20 countries, in Sahelian Africa and in South Asia, have an average per capita food availability of less than 2,000 calories per day.

Cases of regional famines have become rare, but undernourishment (through lack of food) and malnutrition (through a qualitative lack of proteins, vitamins or minerals) are common, especially in the period between harvests when food stocks are virtually exhausted. Although accurate figures are difficult to establish, some idea can be given of the scale.[1] For every ten people:

- One suffers from chronic hunger (around 1,500 calories/day)
- Three suffer from undernourishment (around 2,000 calories/day)
- Three have sufficient energy intake but lack protein.

Any national average conceals wide disparities. Hunger affects first and foremost those who can neither produce nor buy enough food, even though there may be no national shortage: poor farmers and landless agricultural workers, the urban unemployed and shantytown dwellers, and refugees. Women and young children, who are not given priority in the distribution of scarce food, are the most threatened.[2] Some 40 per cent of the world's undernourished are less than ten years old: their physical and mental growth is retarded.

- *Hunger causes 13-18 million deaths per year worldwide (a minimum of 35,000 per day). In developing countries, 60 per cent of deaths among the under-fours are caused by hunger.*[3]

Producing food is not the same as feeding people: the Brazilian paradox

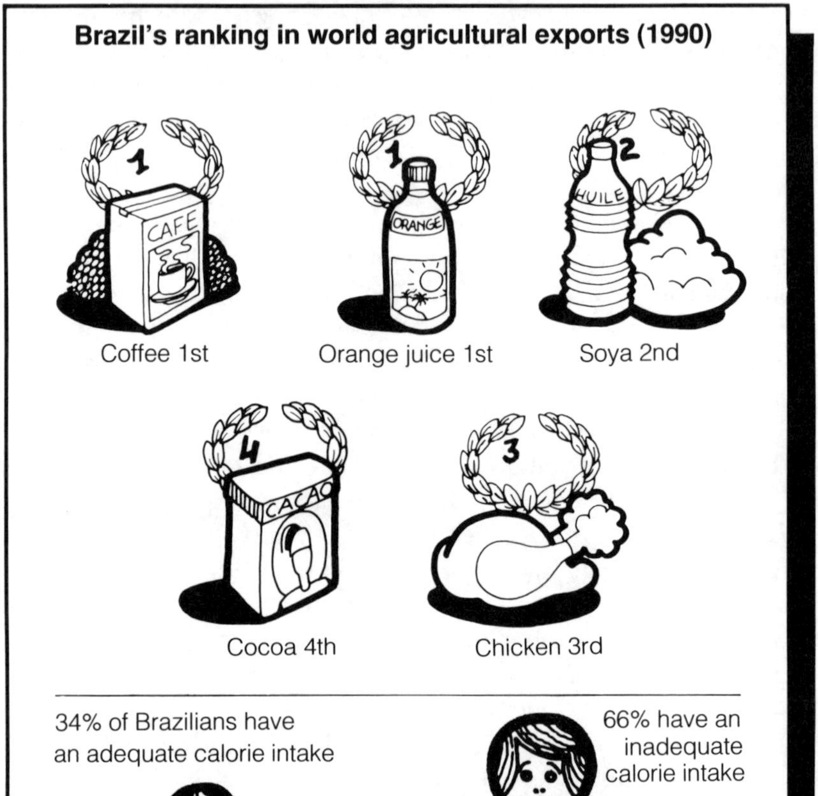

© Delpeuch – Solagral

Brazil is one of those countries, like Malaysia or the Ivory Coast, which have chosen to develop a strong agricultural export capacity. It has become the third largest agricultural exporter in the world. However, two out of every three Brazilians still do not have enough to eat. The issue is not the volume of agricultural production, but the distribution of purchasing power:

- The concentration of income is increasing: the richest 20 per cent of Brazilians have an income 26 times larger than the poorest 20 per cent;
- The distribution of land is unequal: the big landowners defend, with the help of armed militias, their vast underutilised properties, whilst up to 14 million rural workers have no land to farm. In 1988, 18 large landowners owned 18 million hectares (six times the land area of Belgium);[1]
- The quality of diet in Brazil, as in other developing countries, is directly proportional to income. Malnutrition is not so much the lack of food, but rather the lack of earning capacity.[2] Domestic consumption of meat, for instance, was restrained at one time in order to increase exports and help repay the foreign debt.[3]

- *India is the example of a country which has managed a large increase in cereal production as a result of the 'Green Revolution', yet has not eliminated malnutrition. Through lack of income, poor families cannot buy cereal stocks which are then exported.*
- *In Argentina, a major agricultural exporting country, the average calorie availability per capita is higher than recommended levels, but 30 per cent of Argentinian families do not always have enough to eat.*[4]
- *14 per cent of the population of the United States does not have enough to eat.*[5]

Some Third World countries have become large importers of agricultural products

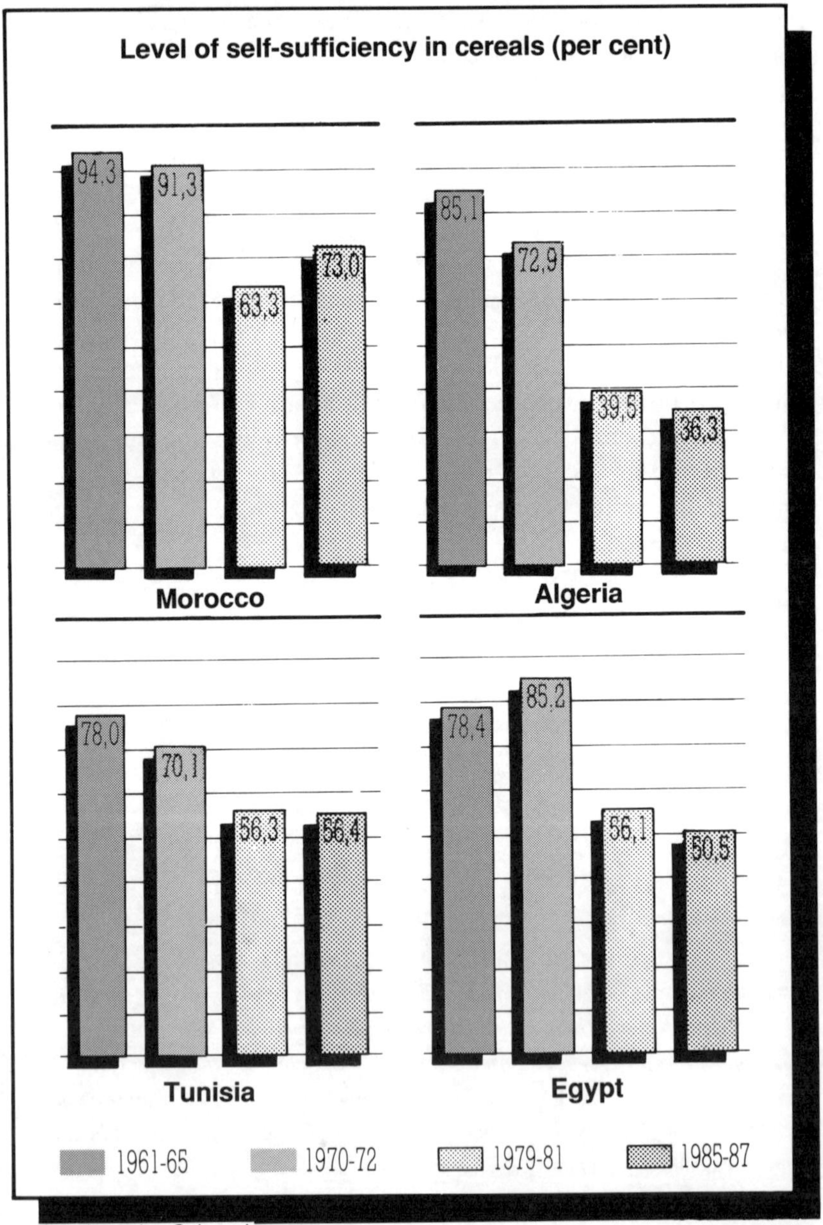

© Delpeuch – Solagral

Source: IAMM

Many Third World countries, having failed to make agriculture a major priority, have had to increase rapidly their food imports: North Africa,[1] the Near East, and oil-exporting countries, for example.

The reasons for this increase in imports of cereals, the staple food of many regions, vary from country to country:

- The stagnation of national cereal production accompanied by a growing population;
- A rise in individual incomes, producing a greater demand for food;
- A tendency towards food diets with a greater proportion of meat, thus requiring more cereals for cattle feed.[2]

So the decline in the level of a country's self-sufficiency in cereals can sometimes result from an improvement in its inhabitants' nutrition.

As long as world cereal prices remain low, importing countries benefit from the price war between exporters. But, in 1988, when when there was an upturn in the cereal market, Egypt's food bill more than doubled in a year.[3] How in the future will Egypt be able to afford the ten million tonnes of cereals which it currently imports every year?

The fall in oil revenues is now revealing the risk taken by those oil-producing countries which neglected their agricultural sector and banked solely on oil to ensure their income. Yet, in the past, several had been exporters of agricultural commodities. Aware of this danger, some countries are now trying, following Algeria's lead, to reverse this trend and give priority once again to agricultural production.

- *In Venezuela, national food production fell from 80 per cent of consumption in 1960 to 50 per cent in 1980. But by limiting foreign exchange available for buying imports, the fall in oil revenues has led to a revitalisation of the domestic market.*
- *Saudi Arabia, although mostly desert, has invested its oil revenue in an ambitious programme of agricultural development and become the sixth largest wheat exporter in the world!*

Africa's growing food dependency

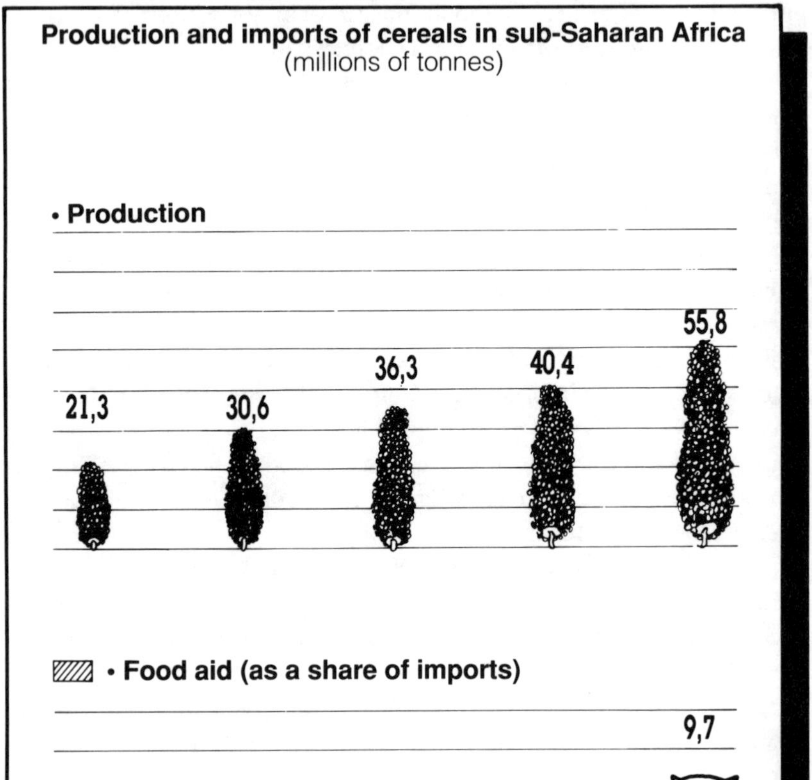

© Delpeuch – Solagral

Food production in Africa is increasing consistently more slowly (1-2 per cent per annum) than population (3 per cent per annum). One African in every five is now dependent on food aid. In the 1980s Africa replaced Asia as the main recipient of international food aid. Yet, in spite of these imports, Africa's per capita food supply has increased by only 0.2 per cent between 1970 and 1990, and remains insufficient to cover minimum needs. Drought exacerbates food shortages which are themselves independent of the vagaries of the climate.[1] In tandem with this increase in imports, Africa's export revenue is falling: there has been a fall in both primary commodity prices and in the volume of agricultural exports. Africa's share of world trade is plummeting.

At the same time, the natural environment is being degraded. The doubling of the population of the Sahel since independence has led to new areas being brought under cultivation[2] and to the end of fallowing as a means of maintaining soil fertility. Yields are low, often 300-500kg per hectare for cereals (United States – 2.5 tonnes/ha; EC – 5.0 tonnes/ha). In fact, as soils degrade, and in the absence of intensive farming, yields tend to fall further still. Due to changes in dietary habits in urban areas, good harvest years do not necessarily mean lower food dependency. City dwellers continue to demand imported rice and wheat, while local millet and sorghum find no buyers, leading to a collapse in prices.[3]

- *In spite of its larger population, sub-Saharan Africa's food imports are smaller than those of North Africa and the Near East, both more populous.*
- *Despite the food crisis, life expectancy in Africa has increased by 30 per cent in 30 years, due mainly to progress in health care.*

A few developed countries account for the bulk of world food trade

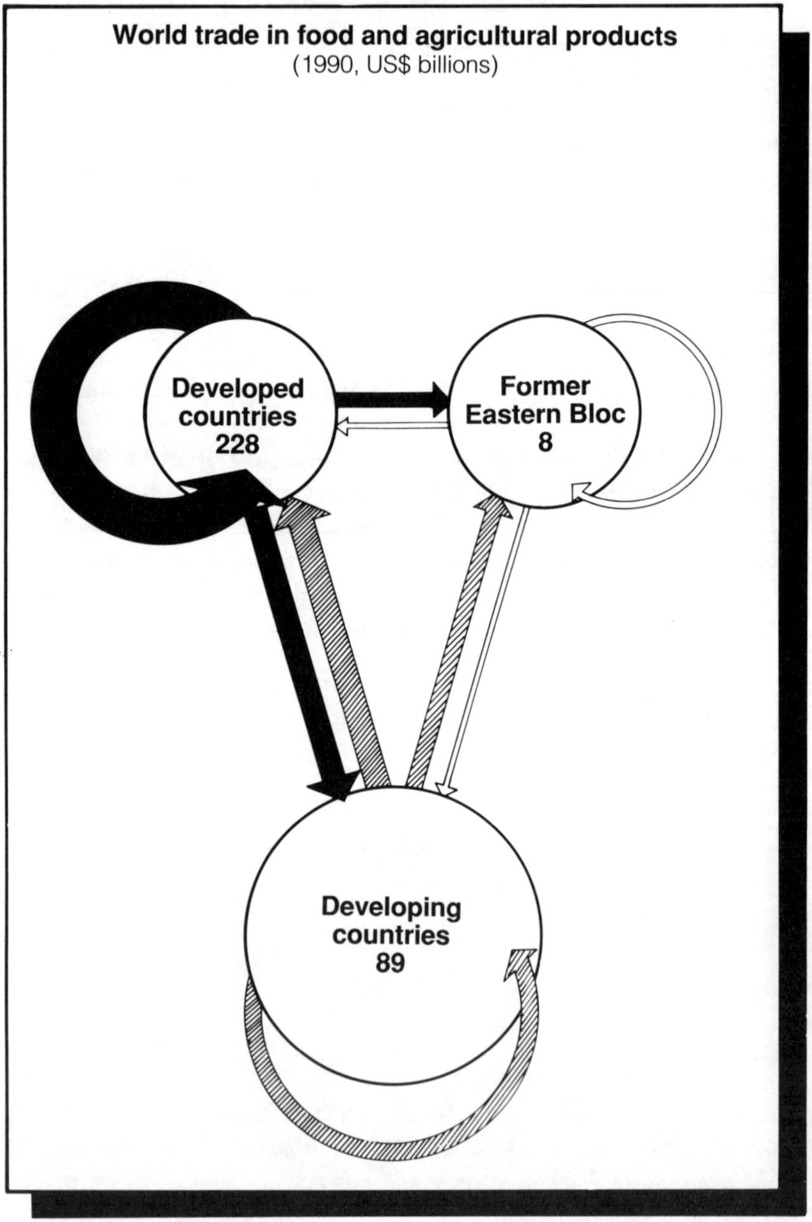

© Delpeuch – Solagral

Source: GATT and J-L Rastoin

The share of food products in world trade is falling, but still accounts for more than 10 per cent. Food is the second most traded category of goods after fuel. However, most of the food produced is consumed in its country of origin. Only one-fifth is exported, with cereals, standardised and easy to transport, the largest by value. The industrialised countries are responsible for most of this trade. Two-thirds of world exports are from 12 countries, two of which, China and Brazil, are classified as developing countries. Similarly, 12 countries account for nearly two-thirds of world food imports.

The developing countries, which supply one-third of all trade in food and agricultural products, have lost some of their export market share in most food products, with only tropical beverages (cocoa, coffee, tea) and some primary agricultural commodities (cotton, jute, rubber, tobacco) maintaining their share. The only increase registered has been for soya and oilseeds, where the world market share of Third World countries has increased as a result of Latin American producers (Brazil and Argentina).[1]

- *The cereal balance of Third World countries has progressively worsened: they produced 105 per cent of their needs in 1936, 96 per cent in 1970, and 91 per cent in 1990. Yet this growing net deficit conceals contradictions:*
  – Oil-producing countries alone account for half of the Third World's agricultural imports;
  – India, the second largest cereal importer in the world in 1966, is now self-sufficient, at least in meeting demand from those who can afford to buy.
- *The UK imported 62 per cent of its cereals in 1909-13, the Netherlands 65 per cent. But both countries were then major economic powers and were not worried about food dependency.[2] The same goes for Japan today.*

The EC: the world's biggest importer of agricultural products

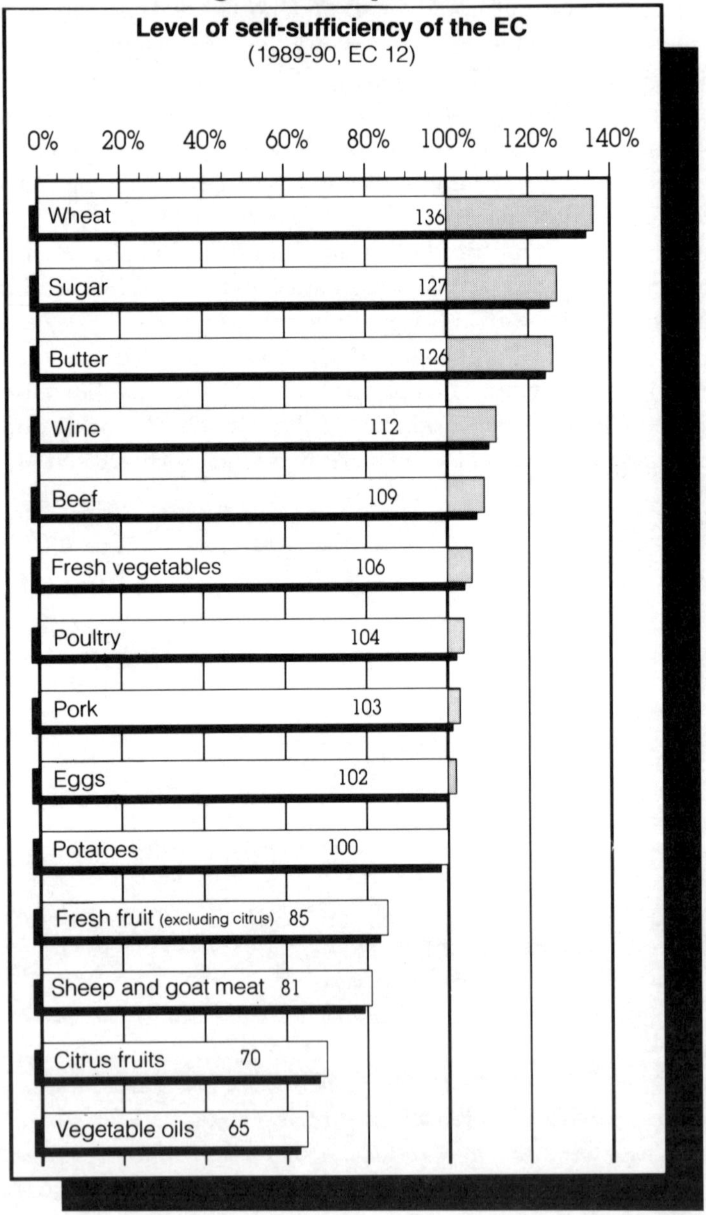

© Delpeuch – Solagral

Source: Eurostat

Founded on the principle of 'Community preference' given to the purchase of agricultural products from member states, the Common Agricultural Policy (CAP) has enabled the EC to become self-sufficient in many products, and to increase its exports rapidly.

The EC has become the world's leading exporter of milk and dairy products, the second largest of sugar and cereals. The sheer volume of its exports and size of its subsidies push down world market prices.

Having become, as a bloc, the largest agricultural exporter in the world for the first time in 1986, the EC 12 remain today the largest importer. The EC's external trade deficit in agricultural products reached Ecu20.8 billion in 1991. The EC not only imports 'tropical' products which cannot be grown – or are little-grown – in Europe (for example, coffee, cocoa, cotton), but also 'temperate' agricultural products which it could produce itself (for example, animal feeds, oilseeds, fruit and vegetables, fish products, timber). Its agricultural trade balance would still be in deficit even if these 'temperate' products alone were taken into account.[1]

- *France is the EC's largest agricultural exporter to non-EC countries, followed by the Netherlands, Germany and the United Kingdom.*
- *The United States is the EC's biggest customer and the EC's biggest supplier of agricultural products. Having accepted, after 1945, that Europe should specialise in livestock and, gradually, in cereals, the United States obtained preferential access to the European market for its soya; this preferential access is also now enjoyed by Brazil and Argentina.*

The EC imports twice as much as it exports to the Third World in agricultural goods

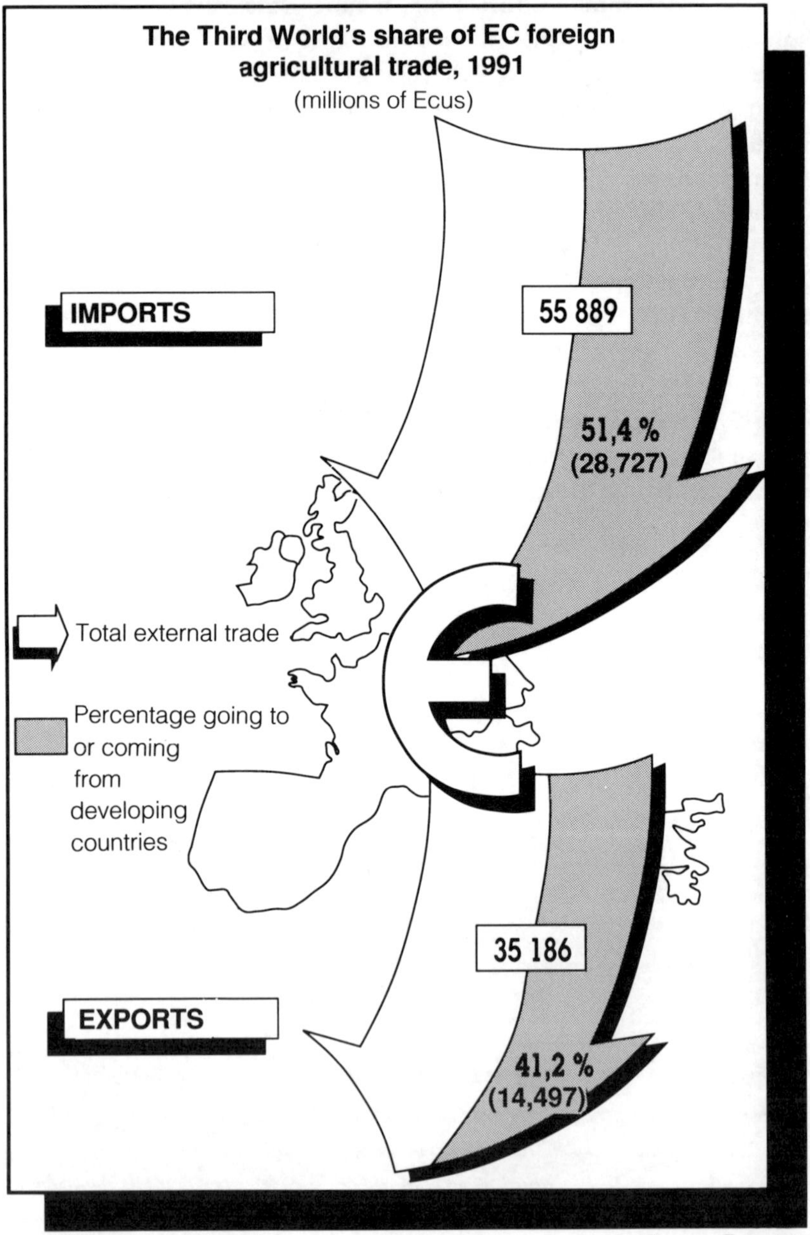

© Delpeuch – Solagral

Source: Eurostat

The Third World supplies half of the EC's agricultural imports. Its share of the European market increased 6 per cent between 1980 and 1990.

By value, the EC imports twice as much as it exports to the Third World in agricultural goods. It takes approximately one-third of the Third World's agricultural exports and has established special mechanisms giving preferential access to Europe through few or no import duties:

- The Lomé Convention, signed with the African, Caribbean and Pacific (ACP) group of countries, which covers mainly tropical products, sugar and meat;
- Agreements with Mediterranean countries (Morocco, Tunisia, Cyprus and Israel) permit access to the EC for their fruit and vegetables (notably citrus fruit, tomatoes and potatoes at times of the year when they do not compete significantly with European producers). But the enlargement of the EC in 1986 to include Spain and Portugal has altered this balance, causing great concern to non-EC Mediterranean countries.[1] Conversely, the EC's agricultural exports to the Third World go primarily to oil-producing countries (half of the total) and to the newly-industrialised countries of Latin America and South-East Asia.

- *Brazil, Argentina, Thailand, China, Ivory Coast, Turkey, Malaysia, Indonesia and Colombia were, in descending order, the main Third World suppliers of agricultural products to the EC in 1991.*
- *When the sales of Thai manioc to the EC came under a 'voluntary export restraint' agreement, the United States took the place of developing countries as the EC's supplier of cereal substitutes. The United States achieved this by increasing its exports of by-products from the food-processing industry, notably corn gluten feed derived from the production of maize starch.*

The United States plays a leading role in world agricultural trade

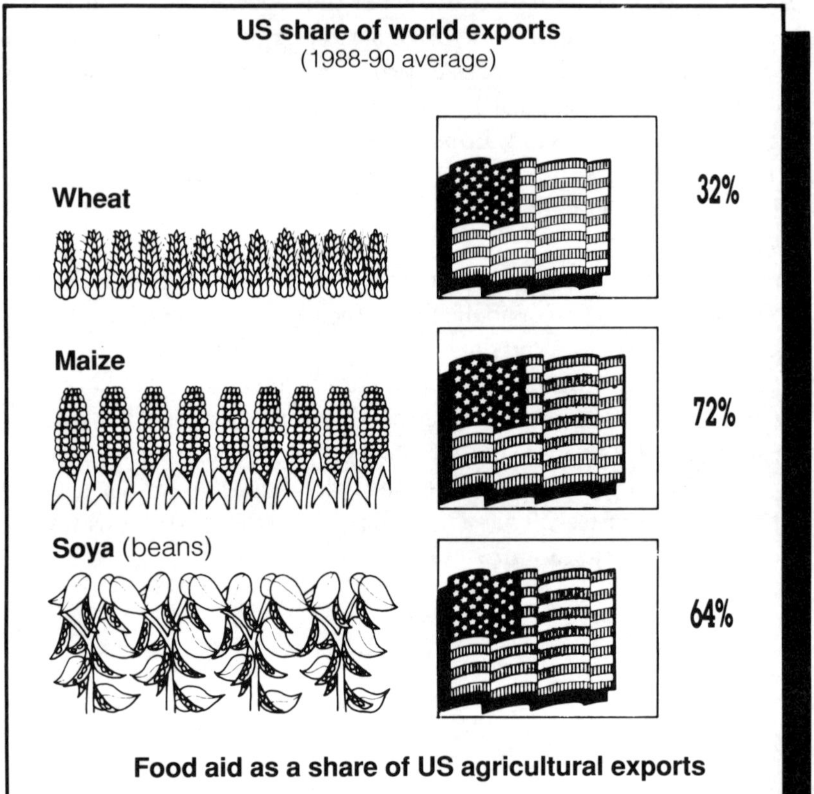

© Delpeuch – Solagral

Source: USDA

The United States still occupies a central place in world agricultural trade, despite a slight decline. More than a third of cereals and oilseeds traded worldwide comes from US farms. In 1970, one hectare in five in the United States was farmed for export;[1] in 1985, two out of five hectares.

The current level of dominance of US agriculture would not have been possible without a strategy of tariff protection and administrative measures aimed at developing export capacity from 1945 (storage subsidies, guaranteed prices, set-aside, the use of livestock production to absorb cereal surpluses). By offering food at low or no cost, the food aid legislation PL 480, passed in 1954, has created new eating habits in the recipient countries and thereby opened up new markets. The United States, as well as being a major agricultural exporter in the world (ranking first for wheat, maize and cotton, and second for rice), is also a major importer of beef, sugar and fish and tropical products. Given its influence on world trade and prices, any change in US domestic agricultural policies has a direct impact on the global food system and the living conditions of farmers around the world.

- *More than half US wheat exports are subsidised via the Export Enhancement Program (EEP).*[2]
- *The decision of the United States in 1986 to subsidise rice exports caused Thailand, the world's leading exporter, to lose substantial revenue because of the subsequent fall in world rice prices.*
- *Marijuana seems to have become the United States' biggest crop, covering 400,000 hectares and generating estimated sales worth US$30 billion.*[3]

With a deficit in animal feeds, Japan is the United States' biggest agricultural customer

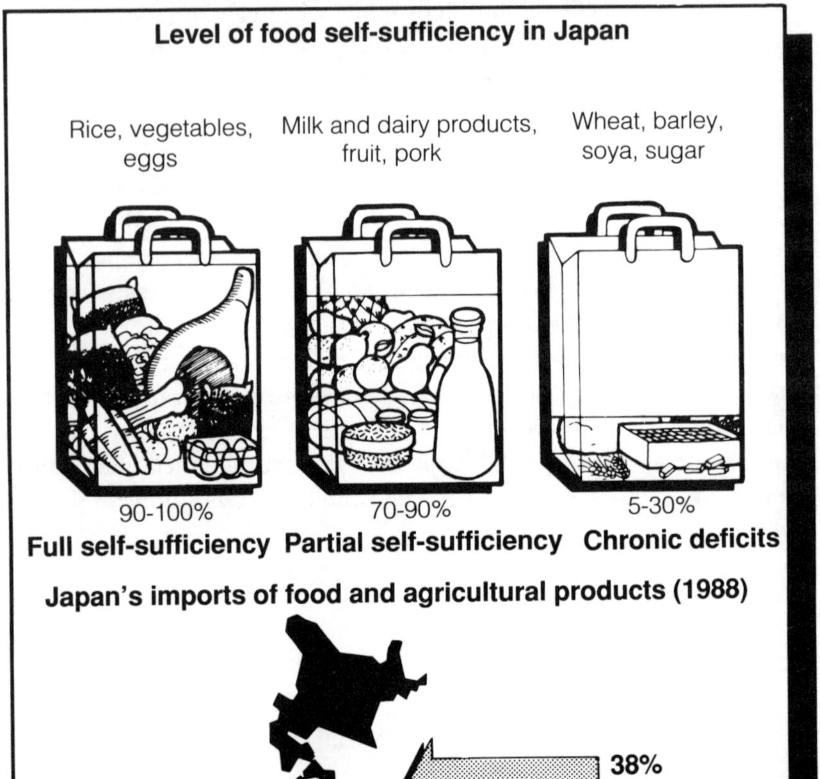

© Delpeuch – Solagral

Having long been on the sidelines of trade between Asia and the West, Japan still employed 50 per cent of its economically active population in agriculture in 1950. Under pressure from the United States, its agriculture and food consumption have been profoundly transformed since the 1960s. The cultivation of irrigated rice, the basis of Japanese civilisation, has enjoyed substantial tariff protection. On the other hand, to meet its demand for animal feed, the country resorted to high imports of maize and soya, mainly from the United States and Latin America. Japan has become the world's biggest buyer of cereals for feed.

Despite the efforts of the authorities to 're-Japanise' food consumption (by slowing down the decline in rice consumption and the rise in meat and bread consumption), state intervention remains essential to sustain the majority of farms, which are too small to compete at world prices.[1] Nowadays, farmers' interests have less and less weight in relation to those of export industries. As a result, the government is gradually yielding to pressure from the United States, which wants to break Japanese agricultural protectionism, particularly in the case of rice, the symbol of Japanese culture.

- *With a huge deficit in its agricultural trade with the Third World, Japan's food and agribusiness companies are very active overseas, especially in South-East Asia.*
- *Because of its topography, Japan has very few cultivated hectares per capita: four times less than China, eight times less than France, 20 times less than the United States.*
- *Although increasing rapidly, Japan's consumption of animal products remains below the average for developed countries. One can be wealthy, eat less than a 'westerner' and live longer while spending less on health!*

Agricultural production is increasingly costly

Purchases abroad of inputs in France

In order to earn a net income of FFr100, a French farmer had to spend

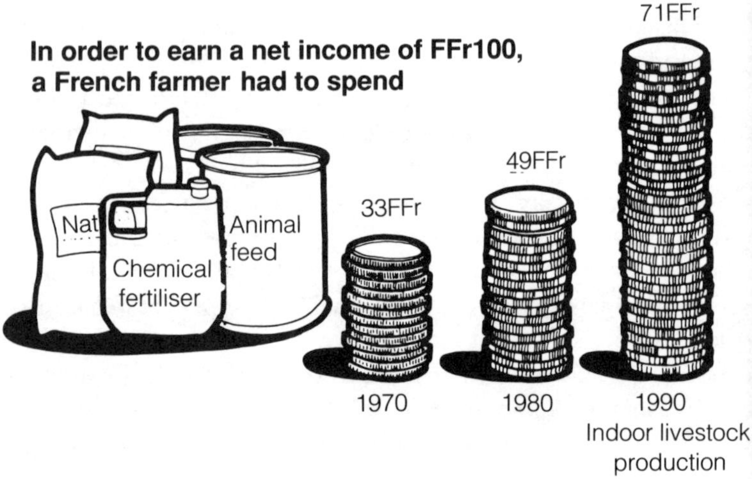

The indebtedness of US farmers is greater than Brazil's foreign debt

Debt in US$ billion, 1989

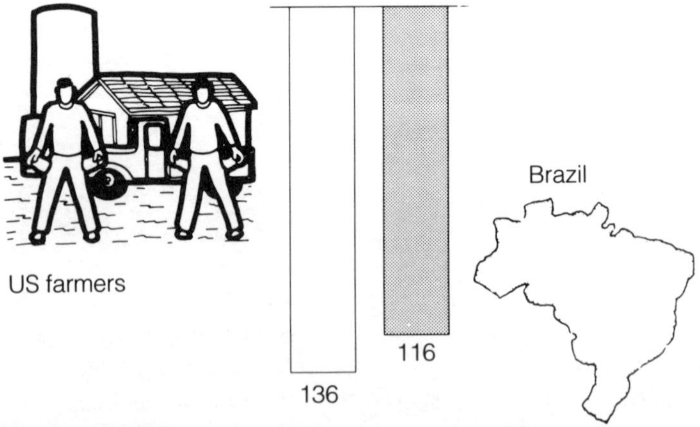

© Delpeuch – Solagral

Source: P.Coulomb & IFRI

Agricultural productivity has grown very rapidly in developed countries since 1945: between 4 and 7 per cent annually in France. This growth is based on an increasing reliance on inputs (seeds, animal feed etc.) bought off the farm. For indoor livestock production, these purchases can reach over 70 per cent of final revenue.

Agriculture has become a heavy industry and highly capital-intensive. In 50 years the capital used per full-time job in farming has risen 15 times in the United States, while at the same time the number of farmers has fallen and land concentration increased.[1] In the Netherlands, the average capital needed by a young farmer exceeds £250,000. With the banks offering credit to farmers in the same spirit as to Third World countries, farmers have had to borrow in order to carry on farming. Many are now heavily indebted, making state support essential to limit social disaster and knock-on bankruptcies in the agricultural supplies industry, especially among farm machinery manufacturers.

- *A third of the EC's farmers are now occupied in other activities off the farm. In the United States, the non-agricultural income of the farming population is now higher than its income from farming.*
- *In agriculture, as in the mining industry, the proportion of fixed costs (land, buildings, livestock etc.) is high compared to variable costs (seeds, fertilisers, chemicals). Working capital per person is comparable with that in the mechanical engineering or steel industries. When prices fall, and in the absence of quotas, the farmer tends to increase production to cover fixed costs, rather than reduce production as classical economic theory would dictate.*[2]

The food industry processes a growing proportion of agricultural products

Proportion of products industrially processed or manufactured for human consumption

United States **95%**

Europe **80%**

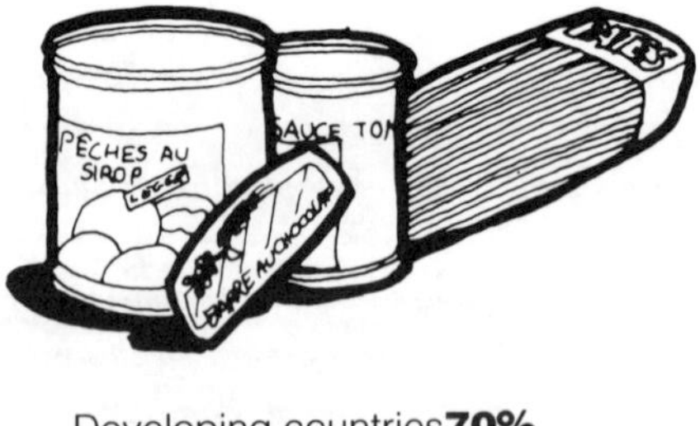

Developing countries **70%**

© Delpeuch – Solagral *Source: J.Chonchol*

As it has intensified, agricultural production has become integrated into a chain of activities both upstream (agricultural machinery, pesticide and fertiliser industries) and downstream (food processing, packaging, transport and distribution). Nowadays, more than 80 per cent of agricultural products undergo some form of industrial processing before reaching the table of the Western consumer. This proportion keeps on growing in line with the demand for easy-to-prepare food: pre-cooked and frozen dishes, ready-to-eat vegetables and so on. The energy used directly by agriculture represents only 2 per cent of the total energy consumption of a country like France. But the food chain as a whole, from fertiliser production through to food consumption, uses 16 to 17 per cent.[1]

The developing countries, as they become more and more closely linked to the international economy, show the same tendency towards a growing integration of agricultural production by the food industry. But their share of the processing remains low. With the exception of Brazil, virtually all the major food brand names are from the North.

- *The active agricultural population in the United Kingdom fell from 4.8 per cent of the total workforce in 1960 to 2.2 per cent in 1990, the lowest in the EC.*
- *The high energy consumption of the agricultural sector can be mainly attributed to the importance of meat-based products and out-of-season vegetables in food consumption patterns in the industrialised countries.*
- *Despite an increase in consumption of processed food, the proportion of income spent on food in the United Kingdom fell to 21 per cent in 1990, compared to 39 per cent in Ireland and 60 per cent for a poor Brazilian family. By the year 2000, the French will spend more on health care (19 per cent) than on feeding themselves (16 per cent).*[2]
- *Vegetables produced in family gardens for consumption by the household still represent 20 to 25 per cent of all vegetable production in France.*
- *In the United States, the value of the wheat contained in a loaf of bread represents no more than 7 per cent of the retail price.*[3]

Around 20 companies account for the bulk of trade in agricultural products

Cereals
Cargill
Continental
Ferruzzi
Louis Dreyfus
Bunge & Born

Bananas
United Brands
Castle and Cooke
Del Monte

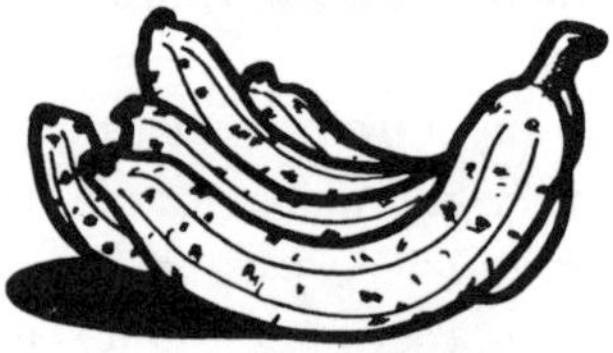

Cocoa
Sucres et denrées
Ed. and F. Man
Cargill

Tea
Unilever
Twinings

Tobacco
Philip Morris
R.J.R. Reynolds
Gulf and Western
American Brand
B.A.T.

© Delpeuch – Solagral

Source: OECD & Ph.Chalmin

World agricultural markets came into being at the end of the 19th century with the fall in transport costs and the introduction of intercontinental communications. These changes led to the development of trading companies with operations beyond their national borders. Around 20 companies are today responsible for the bulk of trade in agricultural products.[1]

The food processing industry is also becoming concentrated in fewer and fewer hands; for several years now, it has registered turnovers hitherto reserved for groups in the chemical, pharmaceutical, finance and oil industries.[2] In the absence of internal expansion, the big food groups grow externally through mergers or buy-outs of other companies. This explains why the giant tobacco companies, worried about the future of the tobacco industry, are looking to diversify their business. Philip Morris bought up General Foods, then Kraft; R.J. Reynolds merged with Nabisco Brands before passing into the hands of investment conglomerate KKR for US$22 billion.[3] This increasing concentration brings with it inflated advertising budgets for products such as confectionery, beer and instant drinks. In the United States, advertising alone accounts for nearly a third of distribution and marketing costs in the food sector. The agrifood industry accounts for 32 per cent of all advertising, whilst its share of total turnover is only 12 per cent.[4]

Only the large transnational groups can therefore afford to invest in the launch of new products and resist the pressure of retailing companies, which create their own brands.

- *In some countries, the state and cooperatives play an important commercial role, parallel to that of the big private traders.*

US companies dominate the world food industry

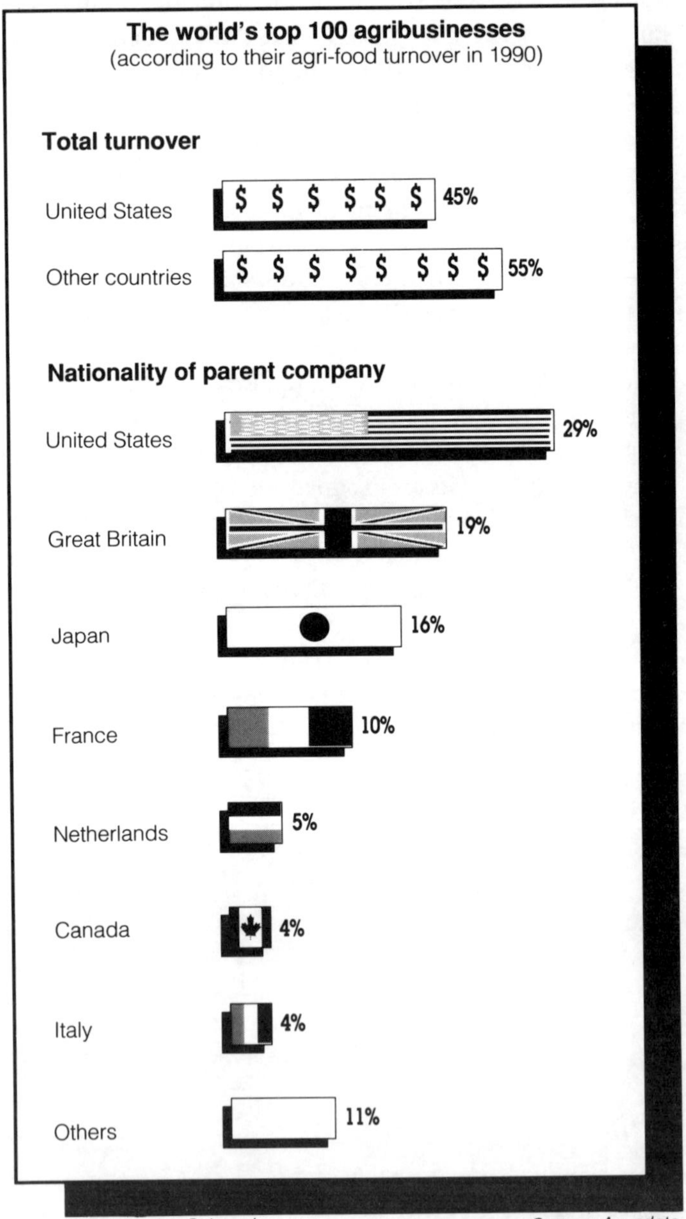

© Delpeuch – Solagral

Source: Agrodata

The list of the 100 biggest agribusinesses in the world illustrates the supremacy of firms originating in the United States. They are the most numerous, together account for almost half the total world turnover, and rank number one in the world in most sectors, for example, cereals, tobacco, chocolate and confectionery, and brewing.[1] US firms are also the most powerful – seven are among the top 12 agribusinesses in the world. Three other countries are also well represented in the list: Japan, France, and, above all, the United Kingdom (seven of the top 30 are British).

The world's top 12 agribusinesses

(1990 turnover, principal activity)

Philip Morris	United States	Multiple products
Cargill	United States	Grains
Nestlé	Switzerland	Beverages
Unilever	UK/Netherlands	Multiple products
Pepsico	United States	Beverages
Coca-Cola	United States	Beverages
Conagra	United States	Multiple products
RJR Nabisco	United States	Tobacco
Grand Metropolitan	United Kingdom	Multiple products
Elders IXL	Australia	Agribusiness
Anheuser Busch	United States	Beer
BSN Group	France	Fresh dairy products

Source: Agrodata

- *Philip Morris and Cargill, numbers one and two in the world respectively, each had a turnover of over US$40 billion in 1990, greater than the GDP of Pakistan.*
- *Size is not necessarily synonymous with competitiveness. Germany's agribusiness sector is the biggest in the EC, yet it has no big firm which counts among Europe's largest.*

A few multinational firms now control the seed industry

Major new entrants in the seed industry (as of 30 June 1992)	
Shell (UK-Netherlands) Petrochemicals	Bought **Nickerson** (UK) (resold to Limagrain), **Sakata** (US), **Abi** (US)
Sandoz (Switzerland) Pharmaceuticals	Bought **Northrup King**, **Zaadunie** (Netherlands) (vegetable and flower seeds), **Hilleshog** (Sweden)
Upjohn (US) Pharmaceuticals	Bought **Asgrow** (US) (vegetables and major crops), **O's Gold Seed** (US)
Ciba-Geigy (Switzerland) Pharmaceuticals	Bought **Funk**, **Steward Seed** (US)
Dow-Elanco (US) Chemicals	Partnership with **Verneuil Semences**
Monsanto (US) Chemicals	Bought **Jacob Hartz** and **Dekalb**'s 'hybrid wheat' programme
ICI (UK) Chemicals	Bought **Garst** (US) (maize seeds), **Sinclair McGill** (UK), **SES** (soya, sugar beet), **Contiseed** (US)
Rhône-Poulenc (France) Chemicals	Partnership with **Seed Tec** (US) (sunflower), bought **Cerès** (France), **Clause** (France). Merged seed interests with **Lafarge-Coppée**
Elf-Aquitaine (France) Petrochemicals	Bought **Rustica** (France), **Dahlgren** (US), **Union Blois**
Lafarge-Coppée (France) Cements (**Orsan**)	Bought **Morran Seeds**, **Claeys-Luck** (second biggest French seed company), then partially resold
Unilever (UK-Netherlands) Food and agribusiness	Bought **PBI** (UK), **Barenburg** (Netherlands) (fodder seeds), then resold
Dupont (US)	Partnership with **DNAP** (US)

© Delpeuch – Solagral *Sources: Agrodata, Grain*

The seed trade started at the beginning of the 19th century for the production of cereals and sugar beet. But it was the perfecting of hybrid maize in the 1920s which marked the real start of the seed industry.[1] Hybrids have a double advantage for the seller: they are automatically protected in that they are both 'non-copiable' for competitors and 'non-reproducible' for farmers who have to buy seeds each year.

Since the beginning of the 1960s, a system of patents has granted marketing monopolies to the 'authors' of new varieties. This new legislation made the selection of non-hybrid varieties economically viable. The way was then opened for the industrialisation of the seed sector, which has been taken one step further by developments in plant biotechnology.

In the United States, then in Europe, large pharmaceutical and petrochemical companies bought up specialised seed houses.[2] Often already established in the agrochemicals sector, these firms sought to sell fertilisers, pesticides and seeds through the same commercial networks.

This concentration of plant breeding in the hands of multinational companies is of concern to developing countries. Now that the international Biodiversity Convention signed in Rio in June 1992 has established the principle of an equitable sharing of 'the benefits arising from the commercial and other uses of genetic resources', developing countries should, in theory, have access to the techniques developed as a result of the use their resources. The Convention, however, represents one more step in the 'commoditisation' of living matter.[3]

The Convention's fatal flaw is that it excludes existing genebank collections – meaning that the four million seed samples purloined from farmers' fields before the Convention was signed remain outside of intergovernmental control. Since this material is readily available, it has the highest commercial value.

Whilst the Convention has legally binding status, its language is often confusing. The text is littered with statements such as 'as far as appropriate' and 'as far as possible', leaving multiple loopholes for abuse and deliberate misinterpretation of the authors' intent. A further serious drawback of the Convention is its bias for bilateral agreements.[4]

- *The use of wild maize resistant to two diseases, collected in Mexico and Guatemala, has made it possible to increase resistance in commonly grown varieties, and to avoid annual losses of US$500 million worldwide.*[5]

World market prices of agricultural commodities are unstable

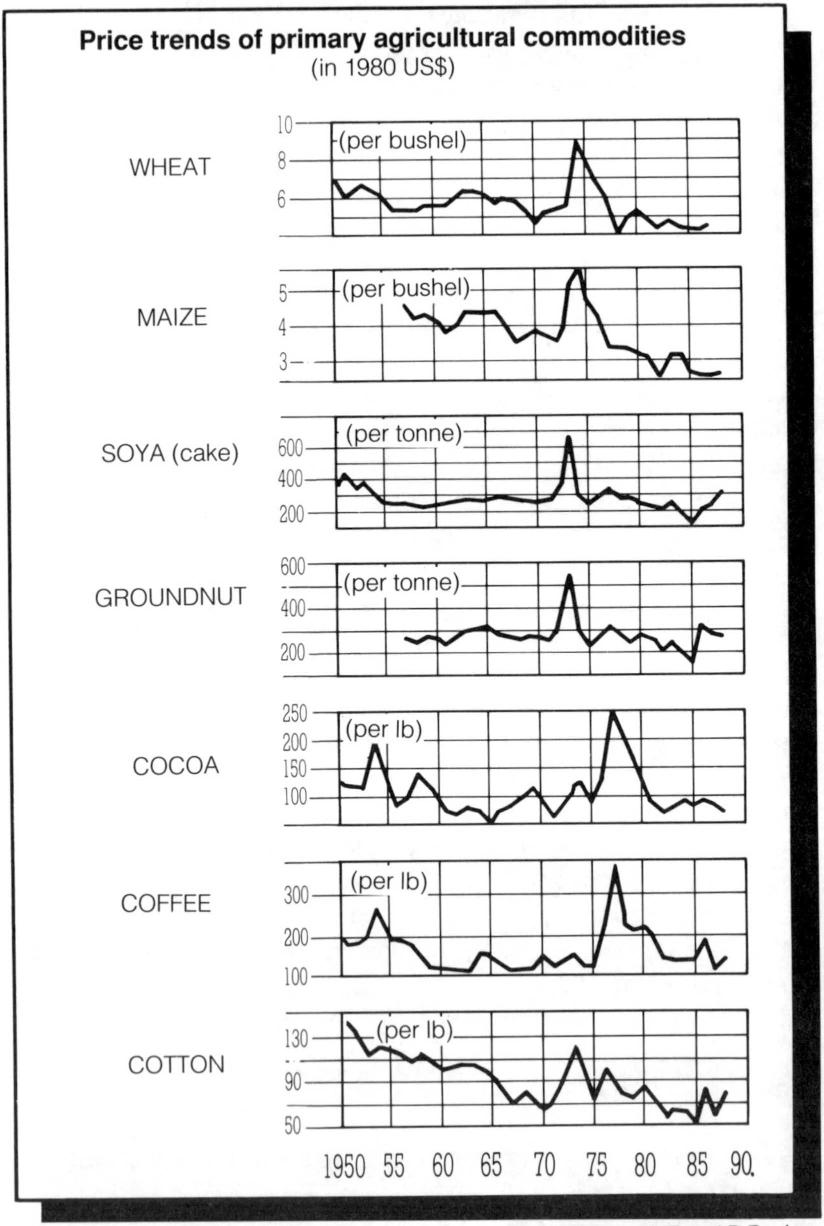

© Delpeuch – Solagral

Source: IMF and B.Daviron

Some agricultural products are mainly intended for consumption in the producing country. Others, such as coffee, cocoa and rubber, are mainly for export on the world market.[1]

Industrialised countries supply the bulk of sugar, cotton and oil seeds. But developing countries are the major suppliers of a number of tropical products, and are all the more dependent on their world price, as two or three of these products often represent most of their export earnings. After years of relative stability, prices of the major agricultural products have undergone significant fluctuations. As well as the vagaries of the climate (frost, drought, flooding) which have always existed, primary agricultural commodity markets have felt the full force of the current instability in the world trading system, in particular fluctuations in dollar exchange and interest rates (the dollar being the currency in which prices are quoted). Recent years have shown that primary producers cannot even expect the downwards trend in prices to increase their sales.[2] Such chronic price instability is all the more serious for developing countries because their economies have little manoeuvring room to adapt. Unlike wheat or maize, cocoa bushes and oil palms do not produce until several years after the initial investment in planting and growth has been made.

- *Nearly half of all LDCs' export revenue is used to pay for the food they import!*[3]

Debt repayments undermine the growth of developing countries

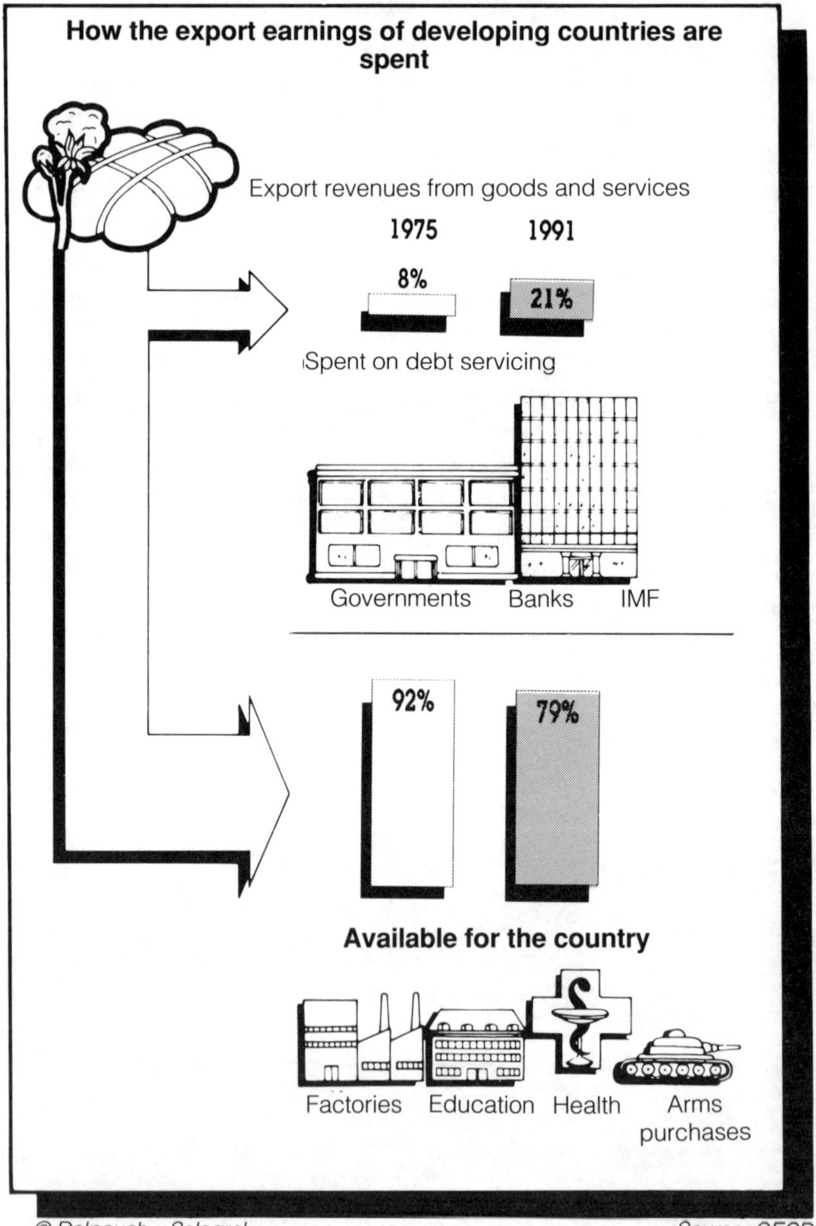

© Delpeuch – Solagral

Source: OECD

The developing countries, in borrowing massively from the banks and from international institutions, thought they would be able to repay the loans both through sales of their primary commodities (the prices of which were rising in the 1970s) and of new products to emerge as a result of this influx of foreign capital. This did not happen, because:

- The necessary size of investments in production (factories, refineries, agricultural projects) was often overestimated and units often did not function at full capacity;
- Other expenditure (administrative, infrastructural), though useful, was not directly productive and did not enable the capital to be repaid.[1]

A country's debt service ratio – the proportion of export earnings which a country must set aside to cover interest payments on its debt – is more significant than the size of its total debt. The toll is a heavy one: around 20 per cent of the Third World's export earnings, up to 30 per cent of Latin America's. For a long time new loans were merely used to pay the interest on rescheduled debts. From 1983 to 1987 the Third World became a net provider of funds to the rich countries, yet this did not mean any reduction in total debt. Since 1989, however, financial transfers have again become positive for the Third World, thanks to an increase in development aid.[2]

- *Western governments are equally responsible for investment projects which are rusting away in Africa. Guarantees from state credit guarantee institutions (COFACE in France, HERMES in Germany, ECGD in the United Kingdom) were given to over-ambitious and often ill-conceived projects.*[3]
- *Third World elites were largely responsible for capital flight. Payments into foreign bank accounts by Mobutu and other leaders from Zaire are estimated at US$4-6 billion, more than the total Zairean debt.*[4] *When ex-President Marcos fled the Philippines, his fortune was put at US$10 billion, one-third of the country's debt.*[5]

Overseas development aid remains below international targets

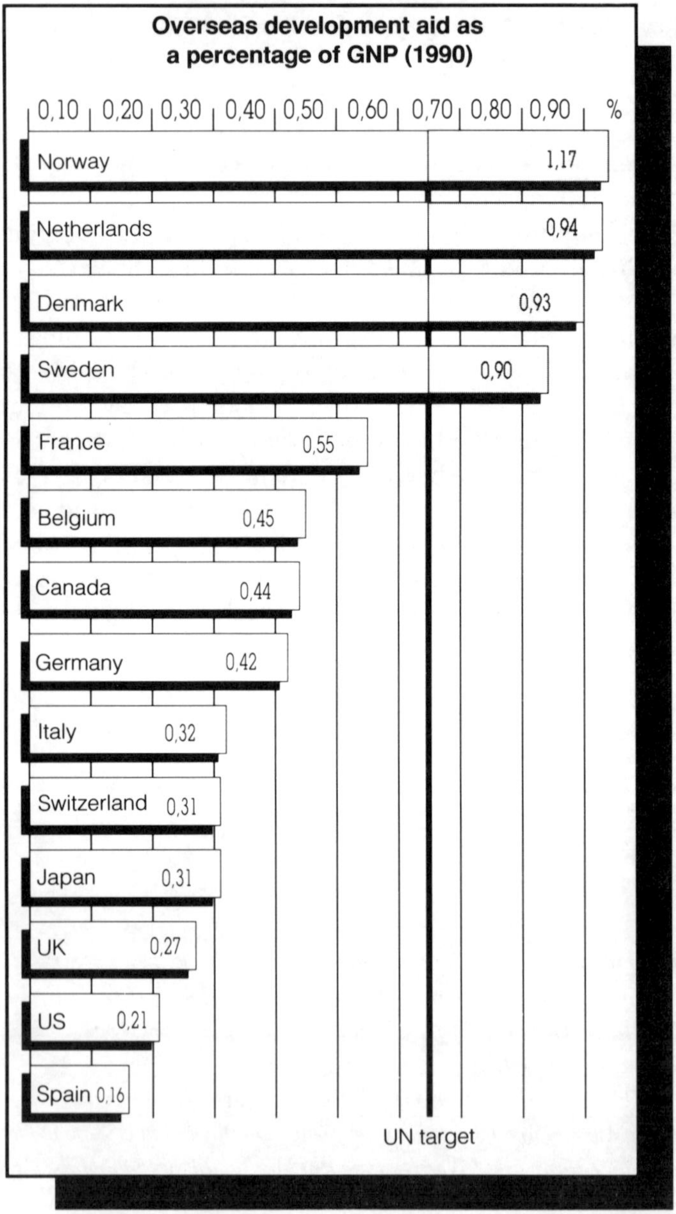

© Delpeuch – Solagral *Source: DAC-OECD*

Development aid was born in the wake of the Second World War with the so-called 'cold war' period. Designed by the United States to prevent the expansion of communism in poor countries, it has since become institutionalised.

Development aid has increased regularly in terms of current US dollars: US$27 billion in 1970, US$58 billion in 1991. But, as a proportion of the GNP of the donor countries, it is virtually at a standstill: for the member countries of the Development Assistance Committee (DAC), which groups the major Western donors and accounts for more than three-quarters of the total volume of world aid, it was 0.34 per cent in 1991 – the same as in 1971. We are still a long way off the commitment made at the UN – and restated at the Rio Summit – to reach 0.7 per cent of GNP.

The volume of aid is not in itself a sufficient criterion of effectiveness. It all depends on how it is spent, and on what priorities determine its use. But it is nevertheless a measure of the importance a country gives to international development issues. Apart from Kuwait and Saudi Arabia, only the countries of Northern Europe reach the UN target of 0.7 per cent. They are also the countries which devote the most resources to raising public awareness on issues of international solidarity. At the other end of the scale, several Western countries (the United States, the United Kingdom and Spain) give a very small proportion of their GNP in overseas development aid.

- *The EC countries together constitute the biggest aid donor, ahead of the United States and Japan. But whilst Japan's aid budget has been increasing spectacularly, that of the United States is stagnating.*
- *The bulk of DAC overseas development aid is paid directly by the donor country to the recipient (bilateral aid). The balance (23 per cent) is channelled through institutions such as FAO, Unicef, WHO (multilateral aid).*
- *Private aid managed by non-governmental organisations, at 8 per cent of the total, remains low compared to government aid. The annual contribution per capita varies dramatically: in 1987, it was roughly £11 in Norway, £6 in Germany, but £1 in France and £0.70p in Japan.*[1]
- *Overseas development aid can also be a big earner. Out of 100 Swiss francs spent on aid, 94 come back to Switzerland in the form of purchases of goods and services.*[2]

The percentage of farmers is falling, but not their overall number

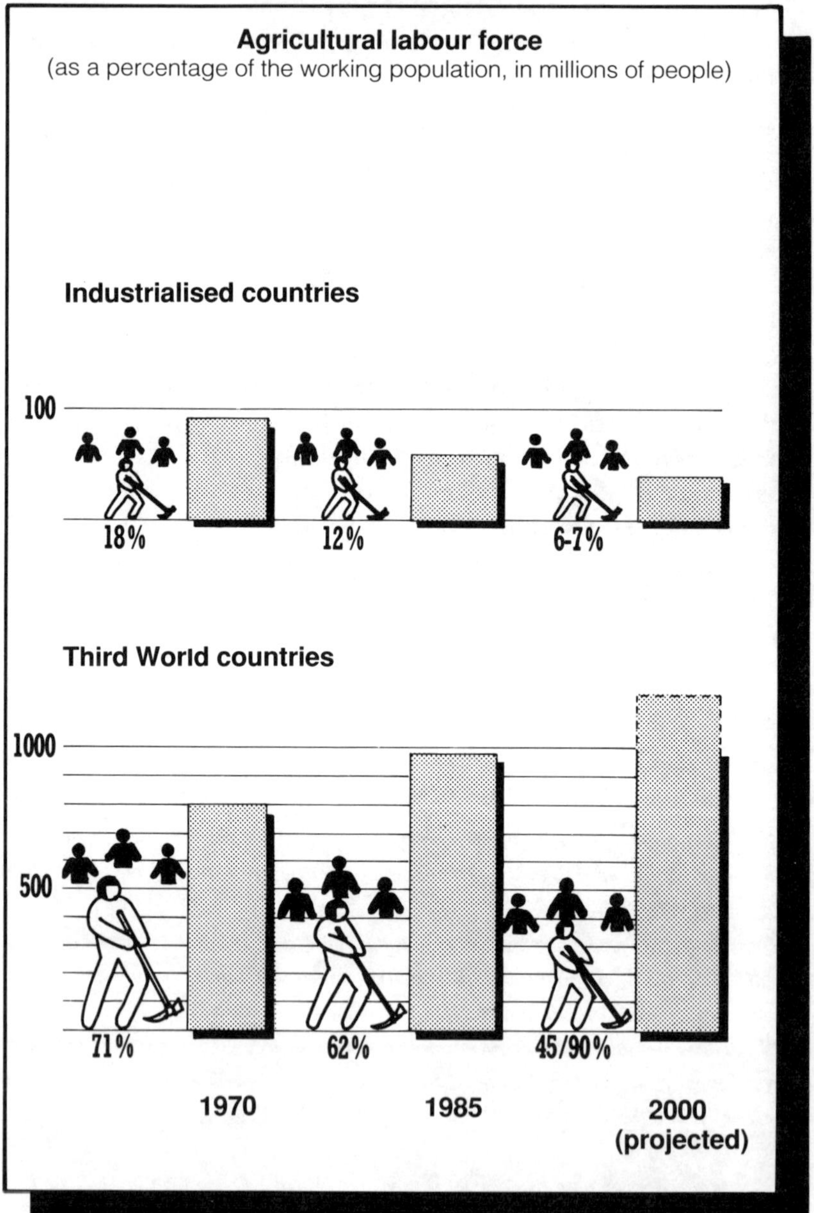

© Delpeuch – Solagral *Source: FAO*

The exodus from the countryside affects all regions of the world, although it takes place at different rates, depending on the country. Unable to make a living from the land, rural people leave to 'make a go of it' in the city. More than 1,000 people arrive every day on the outskirts of Mexico City.[1]

It is in Africa that, proportionally, farmers remain most numerous: 65 per cent of the total workforce (3 per cent in the United States, 7 per cent in the EC). Whilst the percentage of the workforce employed in agriculture is steadily diminishing everywhere in the world, population growth means that the absolute number of farmers continues to rise in developing countries.

For a long time, European farming families arriving in a city could find work in nascent manufacturing industries, then later in developing service industries. Today, there are few jobs available for rural people arriving in the large cities of the Third World. Yet the city retains its appeal with the hope, often dashed for the poor, of making a livelihood more easily than in the countryside.

Left behind in the village, women in certain regions are assuming an increasing share of agricultural work. As young men leave for the city, there is a shortage of hands for certain tasks in the fields. But, beyond a certain threshold of migration, the population in the countryside may no longer be sufficient to maintain the fabric of rural life and look after the environment.

- *The percentage of the labour force in agriculture varies greatly from one developing country to another: 85 per cent in Burkina Faso, 60 per cent in Thailand, 38 per cent in Morocco, 31 per cent in Jamaica and 14 per cent in Chile (1988).*
- *Up to 80 per cent of adult males from some villages in Mali live in France.*
- *Between 1960 and 1990, the European Community lost more than 1,000 agricultural jobs per day!*
- *The timing and speed of the rural exodus have varied according to the country; in 1945 agriculture still employed 30 per cent of the working population in France, but less than 10 per cent in the United Kingdom.*[2]

Women play a crucial role in food production in the Third World

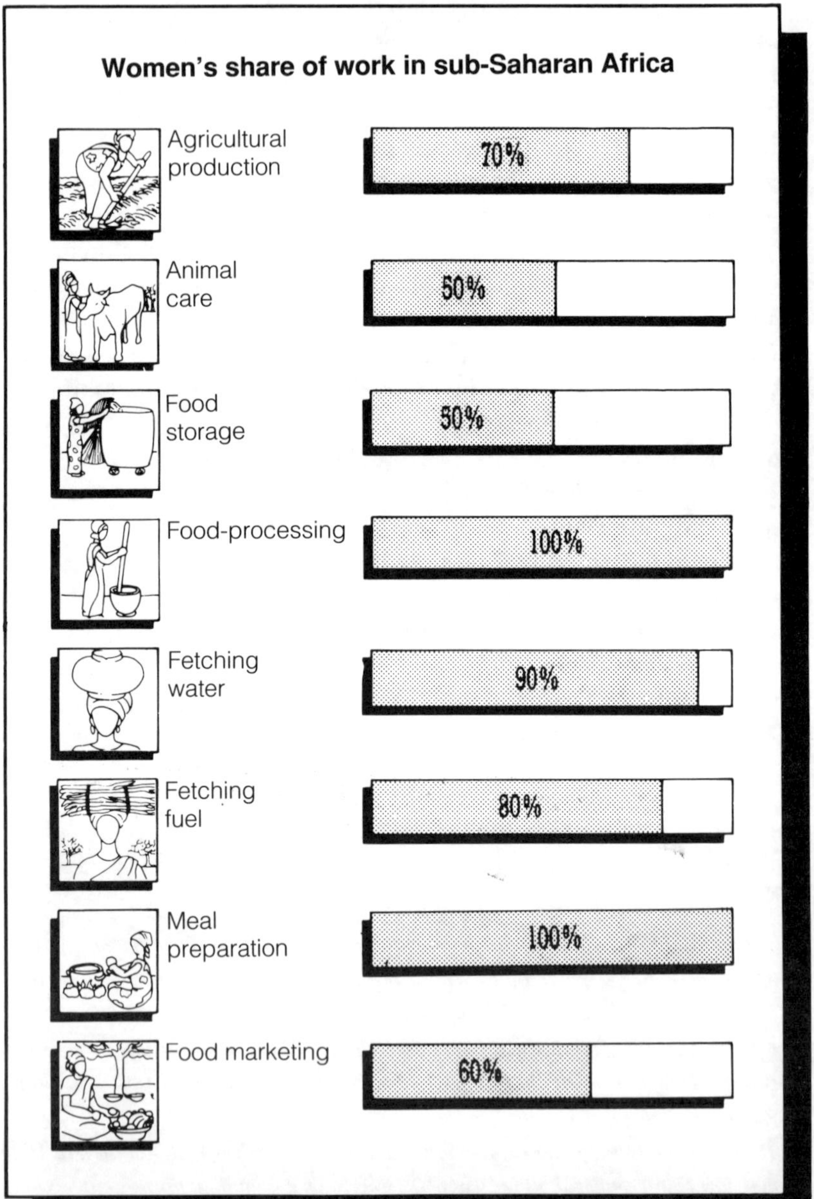

© Delpeuch – Solagral

Source: UN Economic Commission on Africa

Far from being a supplementary activity, women's work in agricultural production and income generation is crucial to the survival of the poorest families. Women do 20 per cent of agricultural work in Latin America, 40 per cent in South and South-East Asia, 60 per cent and more in black Africa. This work is underestimated in statistics, particularly in regions where tradition dictates, as in Arab countries, for example, that the woman 'does not work' (in other words, outside the home).

Reflecting the social importance attached to this work, women are forced to accept very difficult conditions to earn this crucial income: over-work for low yields, more time spent marketing their products once children no longer need supervision, underpaid jobs on plantations and in factories.[1] African women are subjected to the triple burden of:

- Work required in their husbands' fields;
- Subsistence production on their own plots, food processing and selling to earn money;
- Daily household chores and childcare, which are time-consuming[2] and becoming more onerous as a result of desertification.[3]

In contrast to male income, which is often spent on leisure or prestige, female income is generally ploughed back into the family.

LIVERPOOL HOPE UNIVERSITY COLLEGE

- *It is often among the poorest families that the number of single female heads of household is increasing, in Africa as well as in Latin America. In the Sahelian countries, three-quarters of men between 20 and 40 have had to leave for extended periods in search of work.*
- *Women in the Sahel carry on their head more than 300 million litres of water per day.*[4] *Twelve-year-old girls often carry more than 20 kilos of water.*[5]
- *The exploitation of women's work takes place first and foremost within the family. In Cameroon, a man farms on average 1.5 hectares if he has one wife, 2.6 hectares if he has three wives, 3.2 hectares if he has four or more.*[6]

Millions of rural people have no land to farm

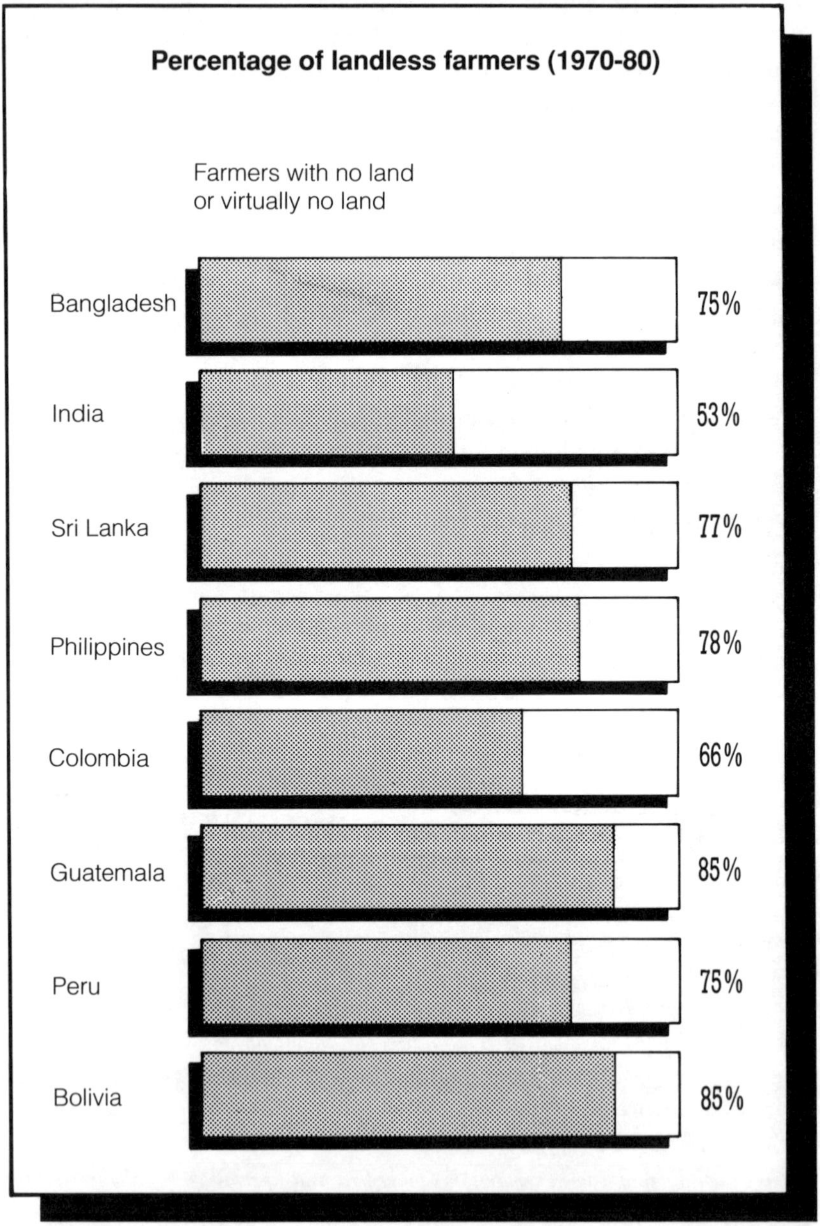

© Delpeuch – Solagral

Source: R.Sinha

A billion people living in the countryside in the Third World have no land, or virtually no land, to work. About two-thirds of them live in India and Bangladesh.[1] Industry and the service sector do not create enough jobs for the existing rural labour force. For rural people, working the land is therefore essential to earning a decent income and living a decent life. Yet, even with two harvests a year, it is difficult for a farming family to secure a subsistence living on a farm of less than half a hectare.

The lack or virtual lack of land for millions of rural people has a variety of causes:

- Relative scarcity of available land, particularly in Asia, where rural population density is high; in India, the 'green revolution' has also enabled the elites to bypass the need for land reform;
- Unequal land distribution with feudal-style land ownership structures. Such a concentration of land is strongest in Latin America;
- Historical relegation to marginal lands. In North Africa, the best lands were confiscated by the French colonists for use by settlers and local elites. In India, the British administration imposed crushing taxes on farmers, condemning them to the loss of their best lands.

Land reform, although often not implemented, or rendered ineffective by pressure from large landowners, continues to be a pressing issue in many Third World countries.

- *Farmers in Bangladesh measure their poverty not against Western lifestyles, but by comparing themselves with the neighbouring wealth of big money-lending landlords.[2] A mere 10 per cent of Bangladeshi landowners control 50 per cent of the land and 50 per cent of the cattle.[3]*

The Western diet is spreading in the Third World from urban centres

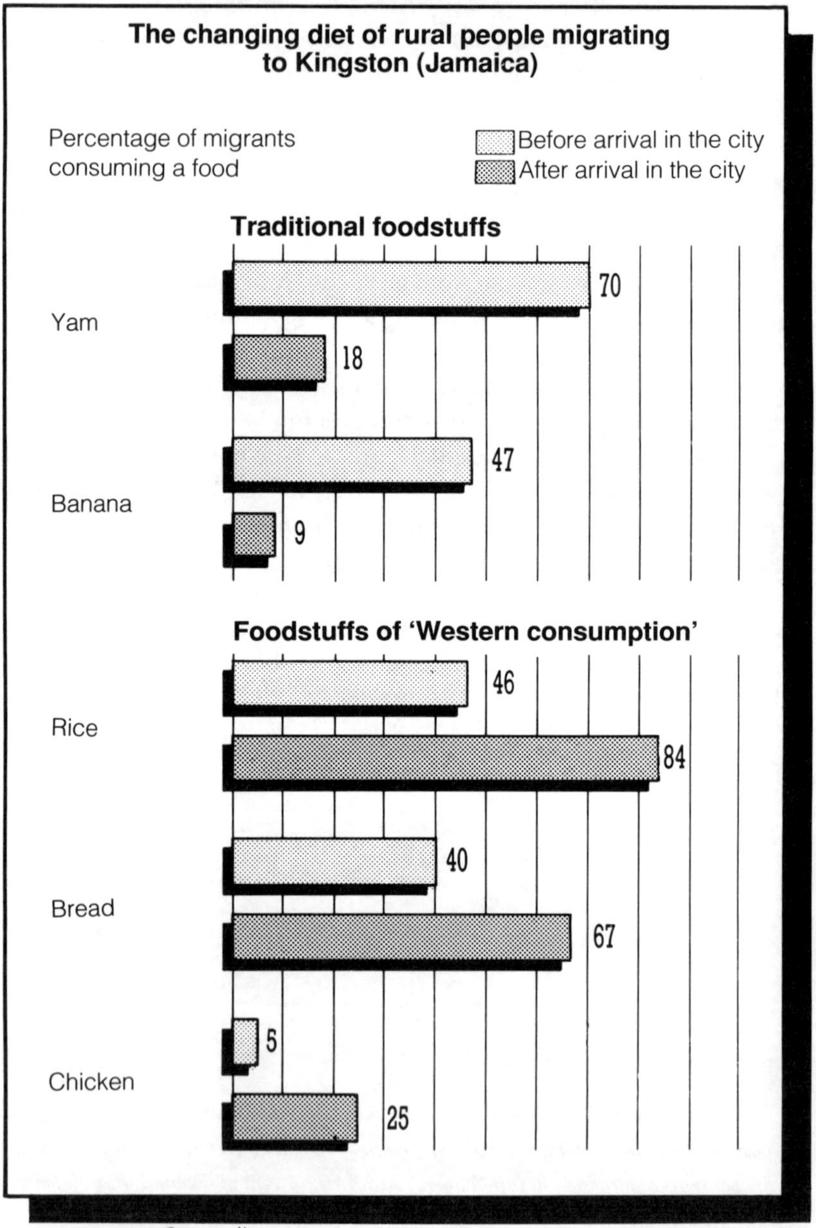

© Delpeuch – Solagral

Source: R.Paquette

Animal products have steadily replaced cereal-based products as the staple food of developed countries. The consumption of meat (notably pork and poultry), dairy products and soft drinks has been rising and the consumption of cereals, fruit, vegetables and alcoholic drinks has been falling. In developing countries, cereals, root crops and tubers still form the staple diet. But traditional crops (millet, sorghum, cassava and sweet potato) are being progressively displaced by rice and wheat. Since 1975, the countries of the South have been importing half of all wheat traded on the world market.[1]

The westernisation of consumption affects the cities first[2] and then spreads to the countryside. While a Senegalese farmer consumes 158kg of millet per year, 19kg of rice and 2kg of wheat, a fellow citizen in Dakar consumes 10kg of millet, 77kg of rice and 33kg of wheat.

Subsidies for the consumption of cheap, imported cereals, the ease of preparation of these foods, advertising, cultural imitation by better-off social strata, and distribution programmes in school canteens are all factors which favour an increase in wheat consumption. Those who can afford it also consume more meat – from animals fed according to the US model, with its high requirements in cereals and soya.

- *The tendency towards a dominant model of food consumption has not destroyed all local consumption habits. Rice is still dominant in Asia and Madagascar, millet and sorghum in Sahelian Africa and wheat in North Africa and the Near East. Within the EC, there are considerable differences in consumption patterns. In 1990 every Italian consumed 110kg of cereals compared with 60kg for each Dutch person. Each Dane consumed 147kg of fresh dairy produce compared with 91kg for each Portuguese citizen.*
- *Bread consumption in Costa Rica has fallen since the arrival of the biggest tortilla factory in the world, set up by a Mexican firm.*[3]

The disappearance of tropical forests is threatening the ecosystem

12 million hectares of tropical forest are disappearing every year

Clearance for agriculture — Collection of firewood — Timber extraction — Extensive cattle-rearing

Average number of hectares deforested to hectares replanted

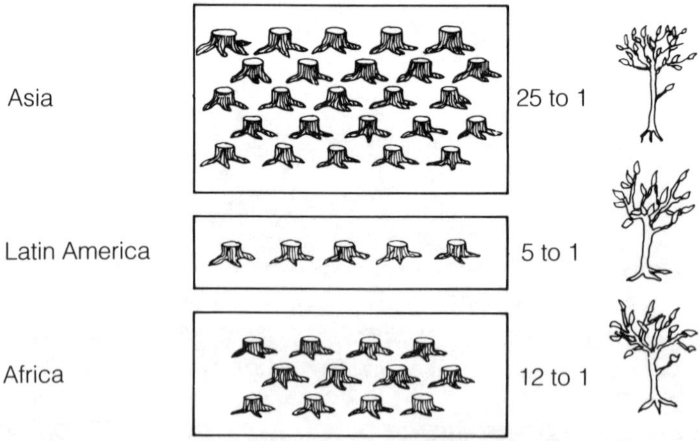

© Delpeuch – Solagral

Source: *Natural History*

An area of tropical forest equivalent to the size of Austria and Switzerland put together is disappearing each year.[1] Some 20 per cent of the forests covering the planet in 1985 will have disappeared by the end of the century.[2]

South-East Asia has replaced Africa as the biggest supplier to the world market of tropical timber, with 85 per cent of shipments.[3] Of 33 countries exporting tropical timber in 1988, only ten will be in a position to continue exporting by the year 2000.[4] Agriculture is a bigger cause of deforestation than timber extraction. Population growth causes people to clear new forests.

The need for fuelwood is another cause of deforestation. In tropical Africa, firewood consumption uses up to 15 times more wood than commercial timber extraction. Cattle-rearing is also a major cause of deforestation. In Central America, 25 per cent of forests lost since 1960 have been destroyed for the ranching of beef cattle, 70 per cent of which were sold to the United States.[5]

Deforestation causes the earth to lose its 'sponge effect', its capacity to absorb then slowly release large amounts of rainwater. Without vegetation cover, the soil becomes more prone to erosion. The disappearance of forests, whose rich natural environments are still little-known, reduces our genetic inheritance and the chances of making use of species of interest to humanity. It also reduces the ability of vegetation to absorb carbon naturally, and in the absence of efforts to reverse the increasing emissions of greenhouse gases, contributes to global warming. At the present rate of reforestation, developing countries will only be able to cover 10 per cent of their requirements at the end of the century. In order to be self-sufficient in just fuelwood by the year 2000, African countries need to increase their present rate of tree planting fifteen-fold.[6]

- *3 million of the world's 4.5 million plant species grow only in the tropical areas.*[7]
- *The Ivory Coast has already lost 90 per cent of its original forest, Nigeria 60 per cent.*[8]

Over-intensive agriculture often neglects the environment

Consumption of ground water in the United States (1985)

© Delpeuch – Solagral

Source: Y.Gazzo

Despite high yields – a guarantee of a cheap and secure food supply – the Western food and farming model cannot be extended to the whole world. It is too demanding in terms of the amount of capital, energy and land needed. A diet rich in meat requires a large amount of land to produce crops for animal consumption. The balance of energy input to energy produced is poor. The model also demands more and more of the environment. Agriculture in the United States is the main user of water, consuming more than three-quarters of the national total.[1] An average of 455 litres is used to produce one egg and 2,000 litres a kilo of wheat.[2] At this rate, the renewal of ground water reserves can no longer be assured.

As well as problems of quantity, there are problems of quality. In several European regions, both in the West and in the East, ground water and surface water are polluted:

- By the over-use of fertilisers and pesticides;
- By the excessive spreading of slurry from indoor animal production onto small areas (one pig produces as much waste as 2.5 people);
- By the ploughing up of marginal (poor or sloping) land, which increases nitrate leaching.[3] The most affected areas are those with a highly intensive agriculture, such as the Netherlands, Brittany, the Po Valley, and southern and eastern England.

Despite progress in recycling techniques, the social costs of this over-intensive model are high: water treatment,[4] soil degradation and the risks to human health are often all borne by the community, which is hesitant about applying to agriculture the 'polluter pays' principle, as in industry.[5]

- *Nitrates take several years to reach the sub-soil. The contamination of aquifers will therefore increase in the future, even if the over-use of fertilisers and slurry was quickly stopped.*

2. UNDERSTANDING THE FACTS

History of agricultural trade

Role of the state

Competition in agricultural markets

The debt burden

Cash crops and subsistence crops

Animal feed, pesticides, seeds: which model of agricultural development?

The failure of collectivisation

The developed countries have replaced the Third World as major agricultural exporters

Four stages of North-South agricultural and food trade

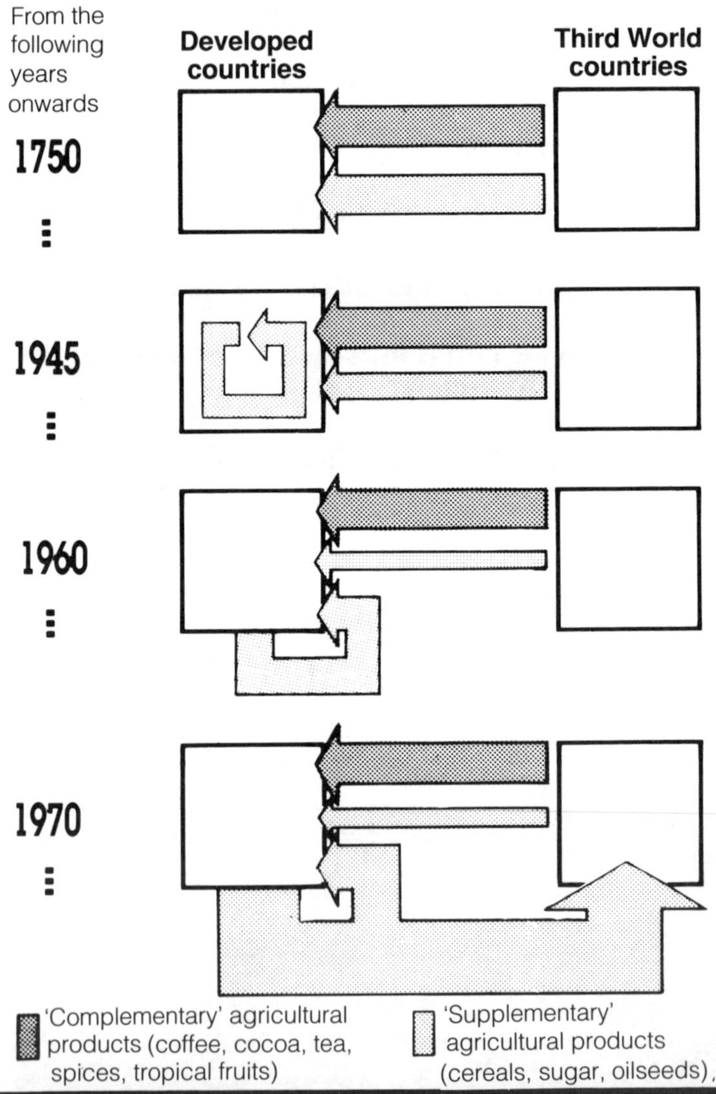

© Delpeuch – Solagral

Source: A. Mounier

While busy developing their own industries, European countries relied for a long time on their colonies in the South to generate their food supplies. These included:

- 'Complementary' products, little grown or not grown in the North, such as coffee, cocoa, tea, spices and tropical fruits;
- 'Supplementary' products, grown in the North but in insufficient quantities, such as cereals, oilseeds, textile fibres and sugar.[1]

After the Second World War, the predominantly South-North flow of agricultural products began to change. Western production of cereals and sugar (as a result of technical progress in agriculture), and of synthetic fibres produced by the petrochemical industry, progressively replaced imports of 'supplementary' products from the Third world.

Growing rapidly from the 1960s onwards, agricultural production in the North entered a third phase: exporting to those industrialised countries which had not yet achieved self-sufficiency. From then on, rich countries bought from rich countries. The flow of agricultural products became increasingly North-North.

The 1970s saw the completion of this reversal of trade flows. Northern agricultural exports to the South grew rapidly, destined for countries which were unable to maintain their market share, except for tropical products and oilseeds (for example, Brazilian and Argentinian soya). The industrialised countries of the North have become major agricultural powers.

Food interdependence between countries is increasing rapidly

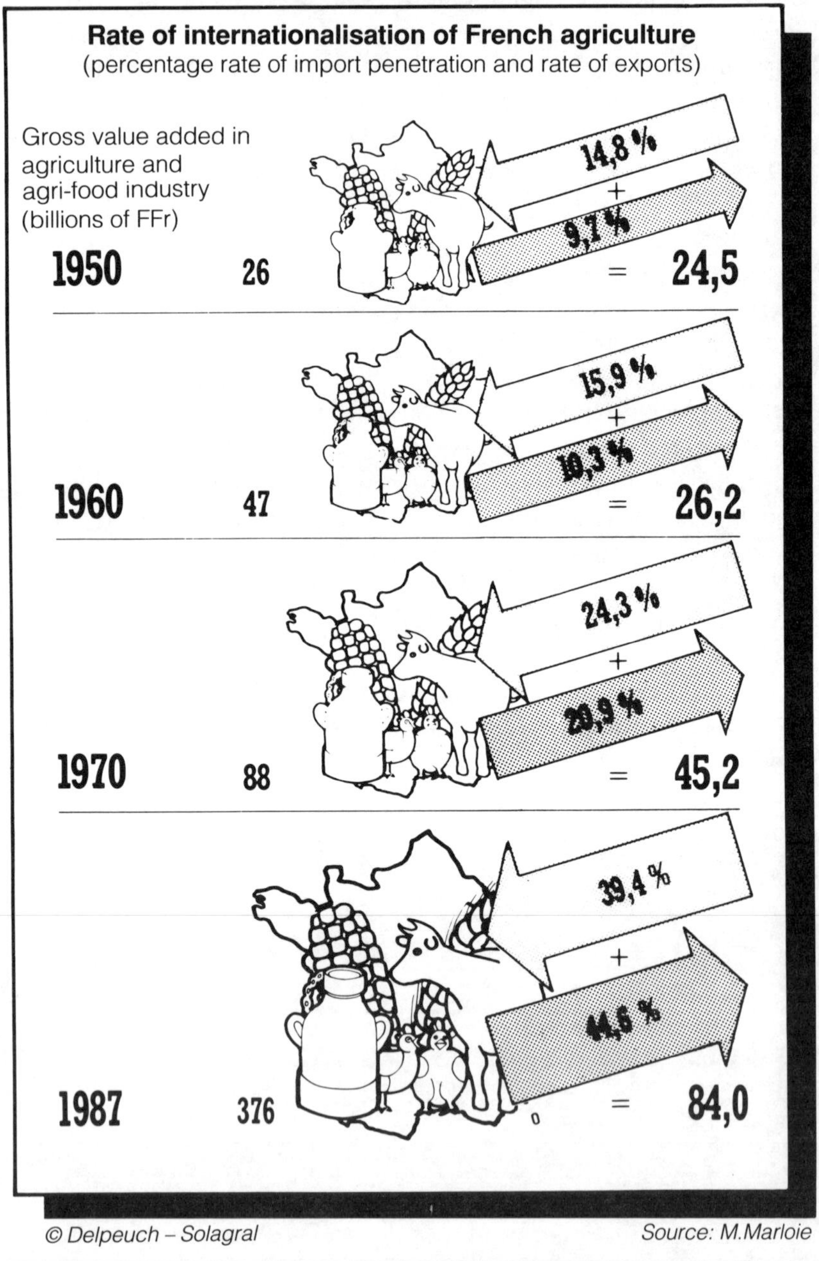

© Delpeuch – Solagral

Source: M.Marloie

The internationalisation of food production started a century ago. Between 1870 and 1890, the import of food products in the West went from virtually zero to 10-25 per cent of production, depending on the country.[1] From 1960 this trend accelerated. World trade in agricultural products grew twice as fast as agricultural production.

In France, home-grown animal feed cakes (sunflower, linseed, rape) represented more than a third of consumption by the livestock industry in 1952, but only 10 per cent 30 years later. At the same time, transatlantic soya imports (from Brazil and the United States) went from 7 per cent to 89 per cent of total consumption, displacing both home-grown cakes and those from former colonies (groundnut, cottonseed, copra and palm kernel).[2] The level of internationalisation of French agriculture therefore rose from 25 per cent in 1950 to 84 per cent in 1987 (as against 40 per cent for the French economy as a whole).[3]

Trade in agricultural food products is gradually drawing developing countries into world trade, rapidly in the case of Brazil and Thailand, more slowly for the land-locked regions of the Sahel or the Andes. Agricultural production is increasingly determined by a world market which sets farmers, whose productivity may differ greatly, in competition with each other. Transnational corporations, which carry out around one-third of world trade in transactions between their own subsidiaries, are a powerful agent in the globalisation of agriculture.

- *The boom in international trade goes back much further than the last century. North African cereals already fed Rome in ancient times. But it was the development in the 19th century of new means of transport (steamships, railways etc.) which made the increased volume and the intensification of long-distance trade possible.*
- *Although a dominant trend, the internationalisation of agriculture has not completely eliminated self-reliance. Three Portuguese farms out of every four marketed less than half their production at the time of Portugal's entry into the EC.*

Decision-making is further and further removed from farmers

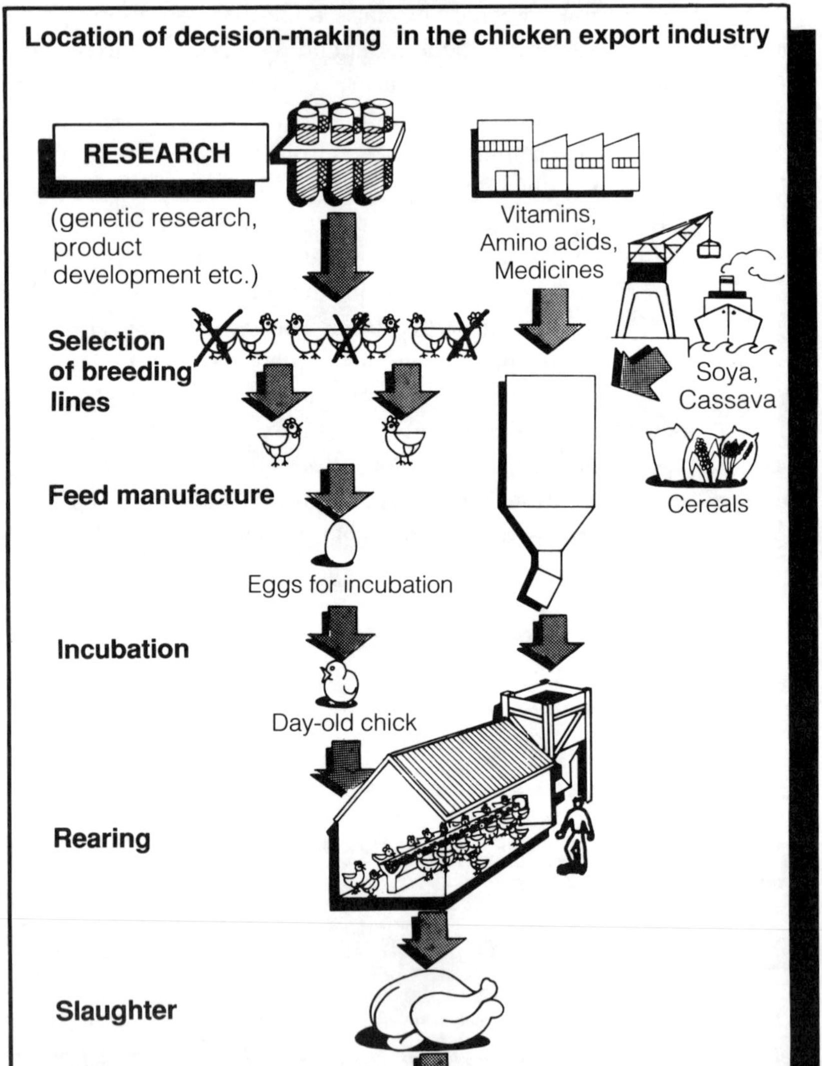

© Delpeuch – Solagral

For centuries, farmers sold the bulk of their products on local markets after initial processing (milled cereals, pressed oils, dried products, for example) or directly to consumers. The development of transport networks progressively expanded this trade. From the end of the 19th century, the concept of a world market price for primary agricultural commodities became a reality in European countries faced with competition from increasing imports from overseas. In 15 years, the domestic price of wheat in France dropped by almost 50 per cent.[1] It was to enable farmers to control the processing and marketing of their produce that agricultural cooperatives were later set up.

Nowadays, the degree of specialisation of agricultural production and the internationalisation of trade have meant a greater distance between producers and decision-makers. In the North as in the South, influence over the direction of agricultural development has shifted from farmers to governments and the big agribusiness and food retailing companies.

- *In the EC, the producers of chicken for export depend for their income on decisions taken by the companies with which they have signed a contract, and on export opportunities. Only companies able to invest in the long term can finance genetic improvements, the development of feeds and customer research overseas. This is why the production levels of chicken-producers exporting to the Middle East fluctuated with developments in the Iran-Iraq conflict and the Gulf War!*
- *In Thailand, cassava producers depend for the sale of their crops on import legislation decided thousands of miles away by European agricultural policy-makers. The decision-making power of the Thai government is limited. When it decided to hold back a proportion of its exports for its national companies, the refusal of the major European importers to buy Thai cassava forced the government to reverse its decision.*[2]

Governments play a crucial role in the organisation of world agricultural trade

How the EC's wheat market is organised

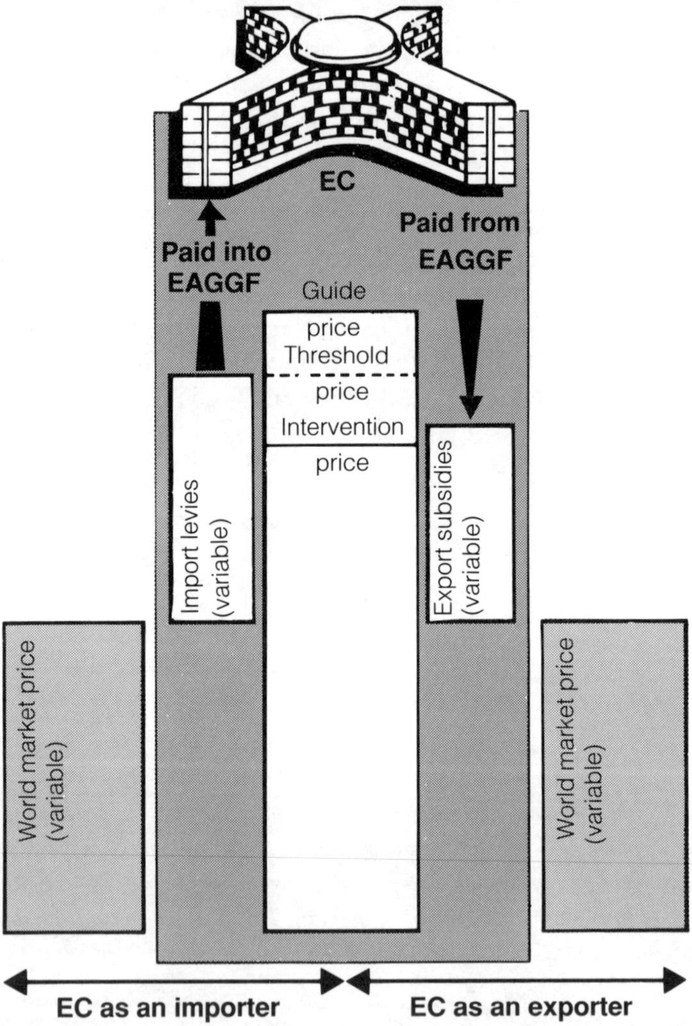

NB: With the reform of the CAP adopted in 1992, the internal EC price should come down to meet the world price.

© Delpeuch – Solagral

In agriculture, governments play a determining role in the organisation of production and international trade. State intervention in agricultural markets goes back to the British colonial companies, the Spanish monopolies, and to the agents of the Chinese emperors. It was reinforced by the crisis of the 1930s, when governments sought to protect national producers from the collapse in agricultural commodity prices. The Second World War further strengthened the role of state bodies established by the major agricultural powers (Canada, France, Australia, Brazil) to control and support their external trade.

Nowadays, the bulk of agricultural products on the world market is managed – directly or indirectly – by state bodies. As partners of the state, international trading companies undertake to fulfil contracts, transport grain and run the risks of price fluctuations. The domestic dimension of EC and US agricultural policy-making is designed to support farmers' incomes, with deficiency payments, storage premiums, investment aids, specific compensatory payments, and so on.

Although declining, the role of the state remains strong in Third World countries, with national marketing boards, agencies with a monopoly over cereal distribution, and food subsidies in urban areas. But the effect on farmers is often the opposite of that in the industrial countries. Whilst Western governments commit funds to guaranteeing their producers' incomes, many Third World countries tax that income in order to feed their budgets. The countryside thus becomes the 'milk cow' of the urban state.

'Green oil' seeks solvent markets

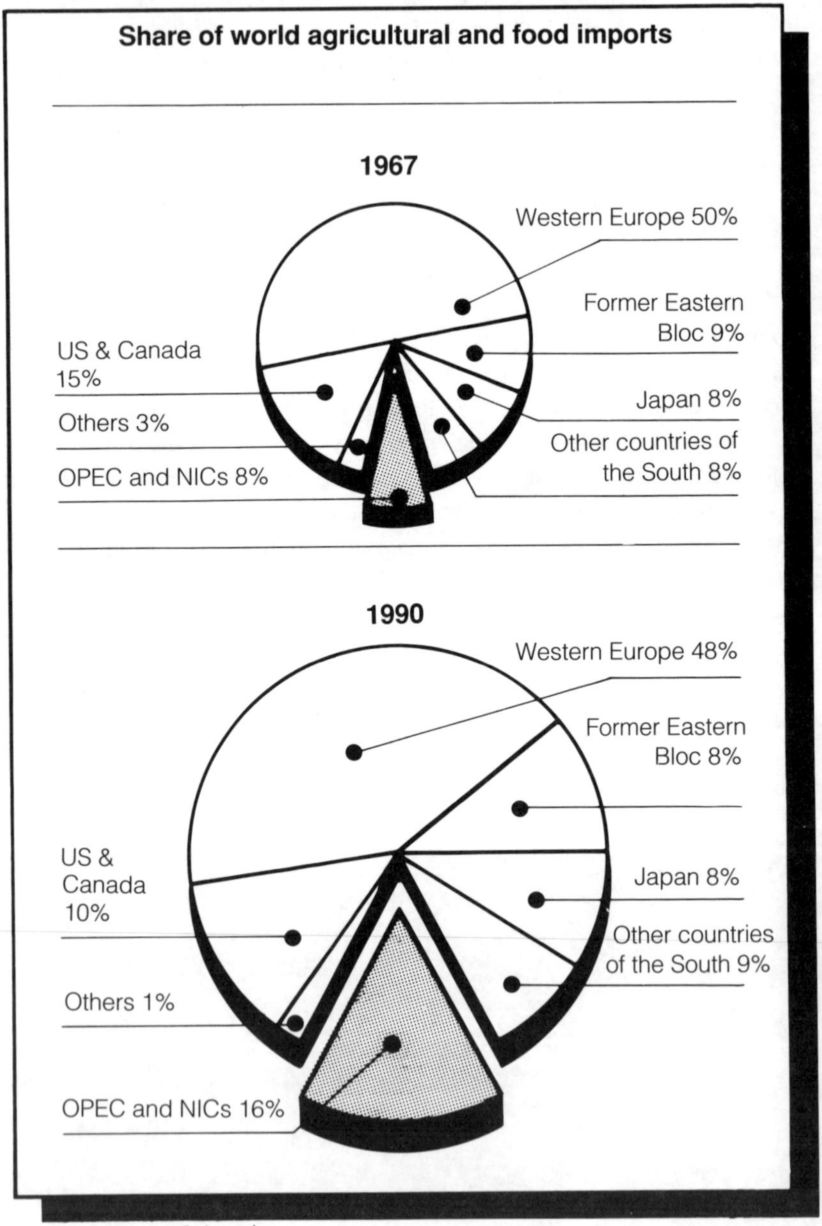

© Delpeuch – Solagral

It was during the 1970s that trade in agricultural and food products underwent an unprecedented boom. Global sales of cereals and oilseeds doubled in ten years. The United States and the EC increased their share of world agricultural exports. Along with the other major agricultural exporters (Canada, Australia, New Zealand, Brazil, Argentina), they drew on sales of agricultural products – their 'green oil' – as a major source of foreign exchange. At the time three types of countries increased their imports:[1]

- The former Soviet Union, which purchased cereals to develop its livestock sector;
- The OPEC countries, whose income went up five-fold after the oil price shock of 1973-74;
- The Newly Industrialised Countries (NICs), which prospered from their exports of manufactured goods.

At the beginning of the 1980s, the trend was reversed. A high level of indebtedness and a fall in income from oil forced certain countries to limit their food imports. Several Asian countries, traditionally importers of agricultural products, were approaching food self-sufficiency (India and China). But the poorest countries, despite an increase in potential need as a result of population growth, could not afford to pay for food imports, given their lack of foreign exchange.[2]

Faced with near-saturated markets, agricultural exporters then waged a subsidy war in an effort to secure market shares for themselves. In the short term, the poorer importing countries took advantage of the situation by buying at low prices on the world market. But there was another side of the coin:

- The agricultural exporters of the South (Thailand in the cases of cassava and maize, Philippines in that of sugar, for example) lost markets because they could not compete with such low prices;
- In food-deficit countries, cheap imports discouraged local agricultural production and increased dependency on the outside world.

International commodity agreements face an impasse

International agreements by commodity in agriculture

Date of first agreement	Initial key mechanism	State of agreement mid-1993
Sugar 1954	Export quotas	Has ceased to function – latest agreement purely administrative
Coffee 1962	Export quotas	Agreement suspended since 1989 – stalled negotiations
Cocoa 1972	Buffer stocks	Producers and consumers unable to agree on the debt owed by producers to the buffer stocks
Rubber 1979	Buffer stocks	The only functioning agreement – due to expire April 1994
Jute 1982	No economic clause	Inoperative agreement due to lack of financial resources
Tropical timber 1985	No economic clause	

© Delpeuch – Solagral *Source: Cyclope*

In theory, the establishment of commodity agreements is simple. In order to stabilise the price of a raw material, the countries concerned reach an agreement to apply at an international level mechanisms which already exist in the developed countries. A quota system allocates a share of the market to each exporting country. A buffer stock, financed jointly by producers and consumers, is intended to maintain prices within an agreed range. If prices fall, supplies are bought into the buffer stock. If prices rise, stocks are sold.

In practice, only the international rubber agreement was working properly in 1993. The others have either not been provided with the economic mechanisms to function or the resources to be implemented. When prices rise, producer countries tend to exceed their quotas in order to boost their income. When prices fall, on the other hand, their need for foreign exchange is so great that they cease to exercise self-restraint in their sales.[1]

The United Nations Conference on Trade and Development (Unctad), set up to limit the impact of market fluctuations, has been progressively displaced by the General Agreement on Tariffs and Trade (Gatt), which, in contrast, seeks to do away with any barrier to the 'free' market.[2] The reduced role of the state in producer countries and the supremacy achieved by the major agribusiness companies have transferred control of the market from the producer to the consumer. This acceptance of market forces as the sole mechanism for fixing prices now makes any attempt at stabilisation around a minimum price futile, given the absence of political will to relaunch the North-South dialogue at a global level.

• *A mechanism originally created under the Lomé Convention, Stabex, allows the EC to compensate falls in income from products sold by the 69 African, Caribbean and Pacific (ACP) countries. Although an exemplary model, Stabex has been unable to prevent a worsening of Africa's position in agricultural markets.*[3]

New technologies increase competition between agricultural raw materials

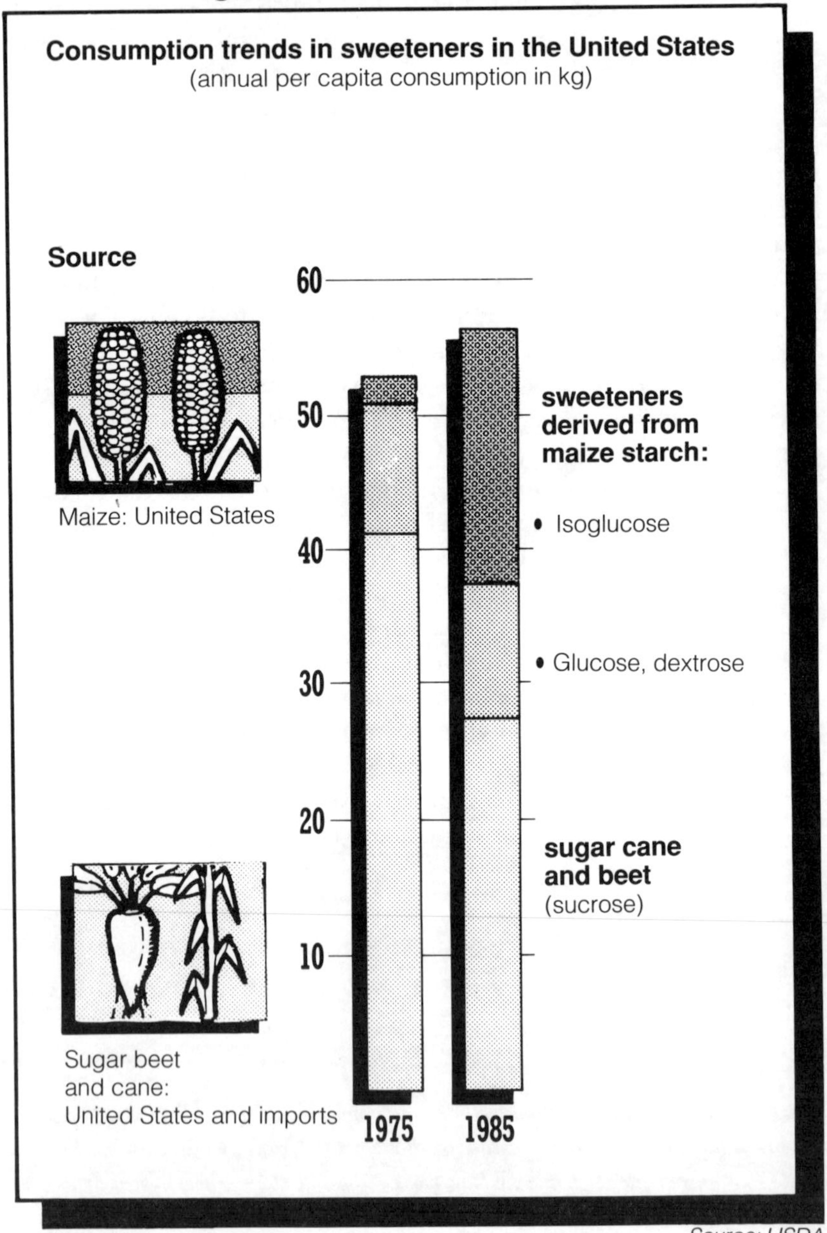

© Delpeuch – Solagral

Source: USDA

The application of new technologies introduces competition between different products intended for the same use. The chemical industry made possible the partial replacement of textile fibres and natural rubber with synthetic products. With developments in the processing of agricultural raw materials (refining and hydrolysis), the possibilities of substitution between agricultural products themselves have increased. Palm oil has replaced soya oil in Asian markets. The different vegetable fats have become interchangeable, enabling large companies to buy at the best price.[1]

Sugar is another example. Since 1985, US consumption of 'traditional' sugar (sucrose) has been lower than that of new sweeteners.[2] Cane and sugar beet are being replaced both by sweeteners extracted from maize (isoglucose, glucose, dextrose) and by synthetic sweeteners such as saccharin and aspartame. The use of isoglucose has grown rapidly in the food industry: in soft drinks, confectionery and biscuit manufacture. It is preferred because of its profitable by-product, corn gluten feed, which is sold for animal feed and exported to the EC where it competes with cereals for fodder. As a result of isoglucose's rise in the United States, Third World countries have seen a sharp reduction in US import quotas for cane sugar.

- *Between 1983 and 1989 the Philippines' US sugar quota fell from 400,000 to 180,000 tonnes.[3] Thrown out of work by the loss of this market, cane workers on the sugar island of Negros asked Coca-Cola and Pepsi-Cola to assume their 'social responsibility'. These two companies, having decided to meet 100 per cent of their requirements for soft drinks with maize isoglucose, have contributed to the dramatic fall in US cane sugar imports. The sugar workers' union asked them in the spring of 1987 for financial help to convert the land to subsistence production.[4] Without success...*

Multinational companies profit from the differences between countries

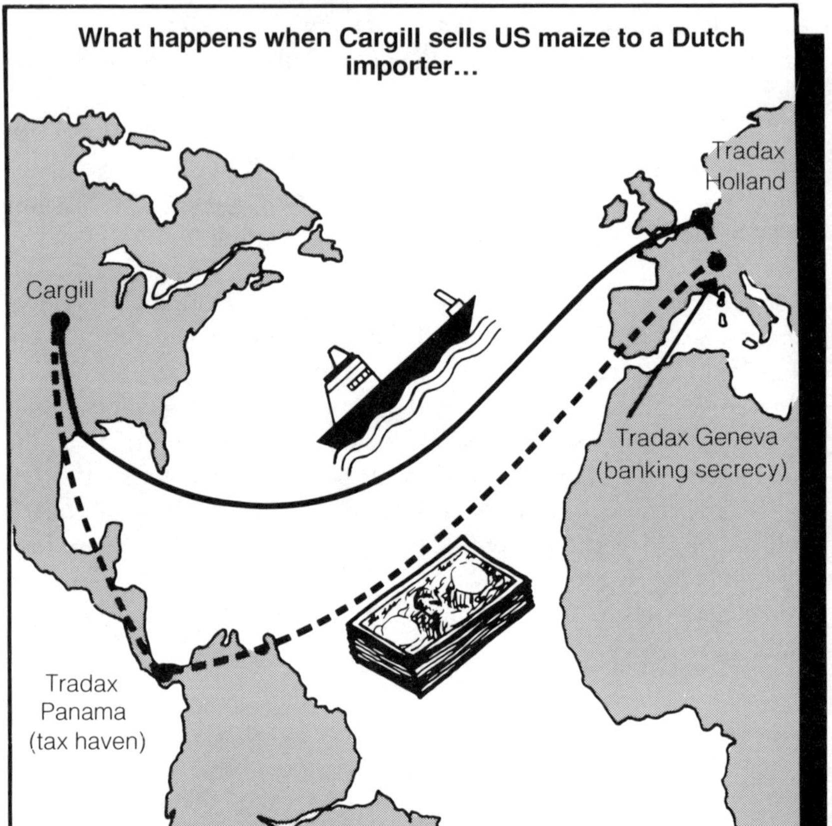

© Delpeuch – Solagral

Source: D.Morgan

The big agribusiness companies now rarely involve themselves at the production stage. They usually prefer to concentrate on the processing and marketing of products where the value added is higher.[1] Very few of them still own outright big plantations in the Third World. But as indispensable intermediaries in international trade and in inter-governmental agreements, they retain indirect control over agricultural producers whom they set in competition with their counterparts in other parts of the world.

The share of world trade carried out within multinational companies between their subsidiaries was estimated by Unctad to be 30 per cent in 1990. This captive trade enables them to manipulate prices during the course of internal transfers. Through systems of over- or under-pricing for services or products sold, the profits are repatriated from the subsidiaries to the multinational's home country or, increasingly, to a holding company established in a tax haven.[2]

The multinationals (some of which belong to the public sector) also take advantage of their bases in several countries to circumvent national laws such as exchange controls, profit taxes, trade restrictions in the wake of an embargo and so on. With the world economy in crisis, the influence of multinationals has become greater; and governments compete with each other to attract them to their countries.[3] The growing concentration of companies, their diversification into services, and their close links with the banks, increase their clout with governments and their ability to take advantage of differences in legislation.

The agricultural systems of developing countries are penalised

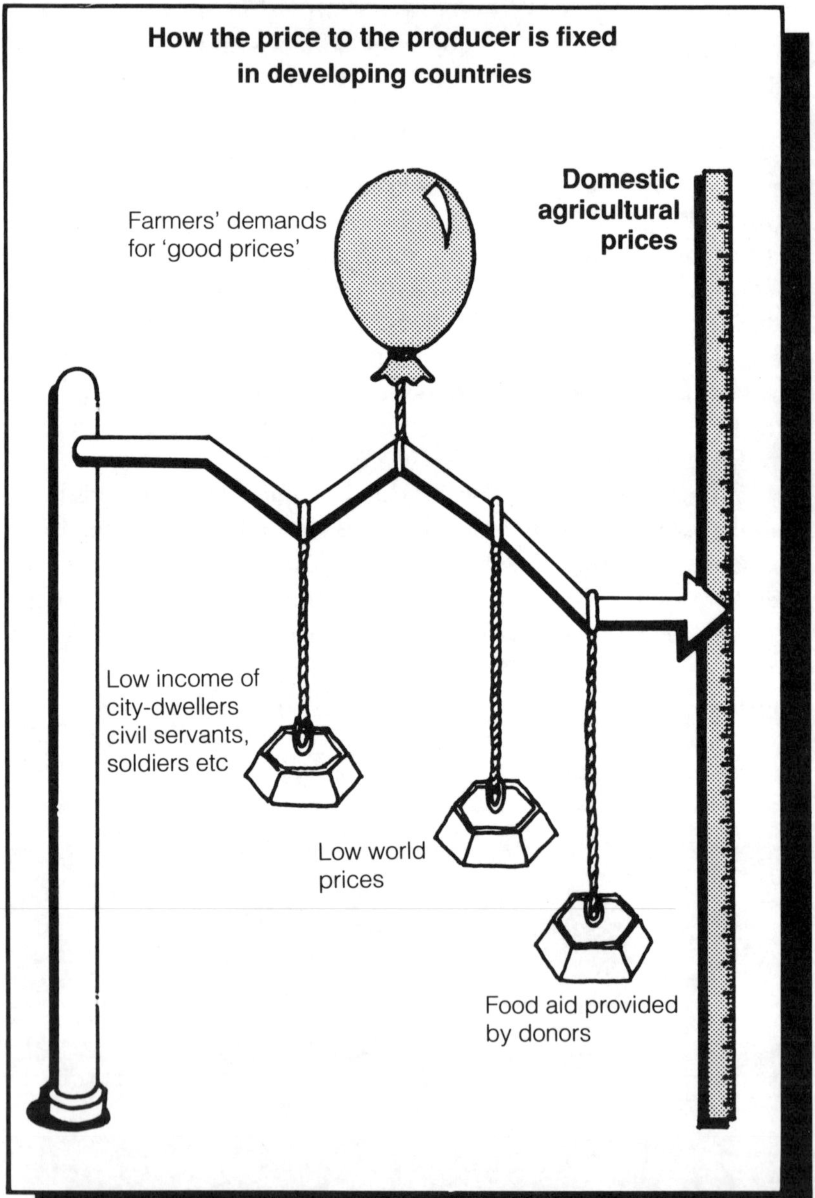

© Delpeuch – Solagral

Source: E.Pisani

Developing country governments, when they set the agricultural prices to be paid to their producers, are subject to a single source of upward pressure: that of farmers wanting a 'good' price. However, farmers are often weakly organised, live a long way from capital cities and are little inclined to market their products.

At the same time, three factors tend to push down domestic agricultural prices: the low income levels of those on whom the government's political stability (civil servants, army, city dwellers) depends; relatively low world prices, further depressed by subsidised Western cereal exports; and the possibility of resorting to food aid.[1]

Export crop producers are no better off. In Africa, their products are taxed by governments through marketing boards set up in colonial times. In the early 1980s the price paid to the producer for cocoa was less than half that paid at the Ghanaian border, and one-third in the case of coffee leaving Togo.[2] This tax on agricultural incomes levied by the boards has tended, on the whole, to increase since independence.[3]

Taxing export crops may have had laudable objectives, such as financing the development of subsistence production, or encouraging agro-industries by penalising the export of unprocessed products. But these taxes have often been used to invest in other sectors of the economy instead (industry, infrastructure), and to support the activities of the public sector. Excessive taxation has:

- Encouraged smuggling into neighbouring countries;
- Discouraged cash crops and lost the country export revenue; Tanzanian cotton, cashew nut and tobacco producers stopped growing these crops during the 1970s and instead planted maize, which commanded a higher price on the black market.

African governments have long discouraged their farmers by favouring imports

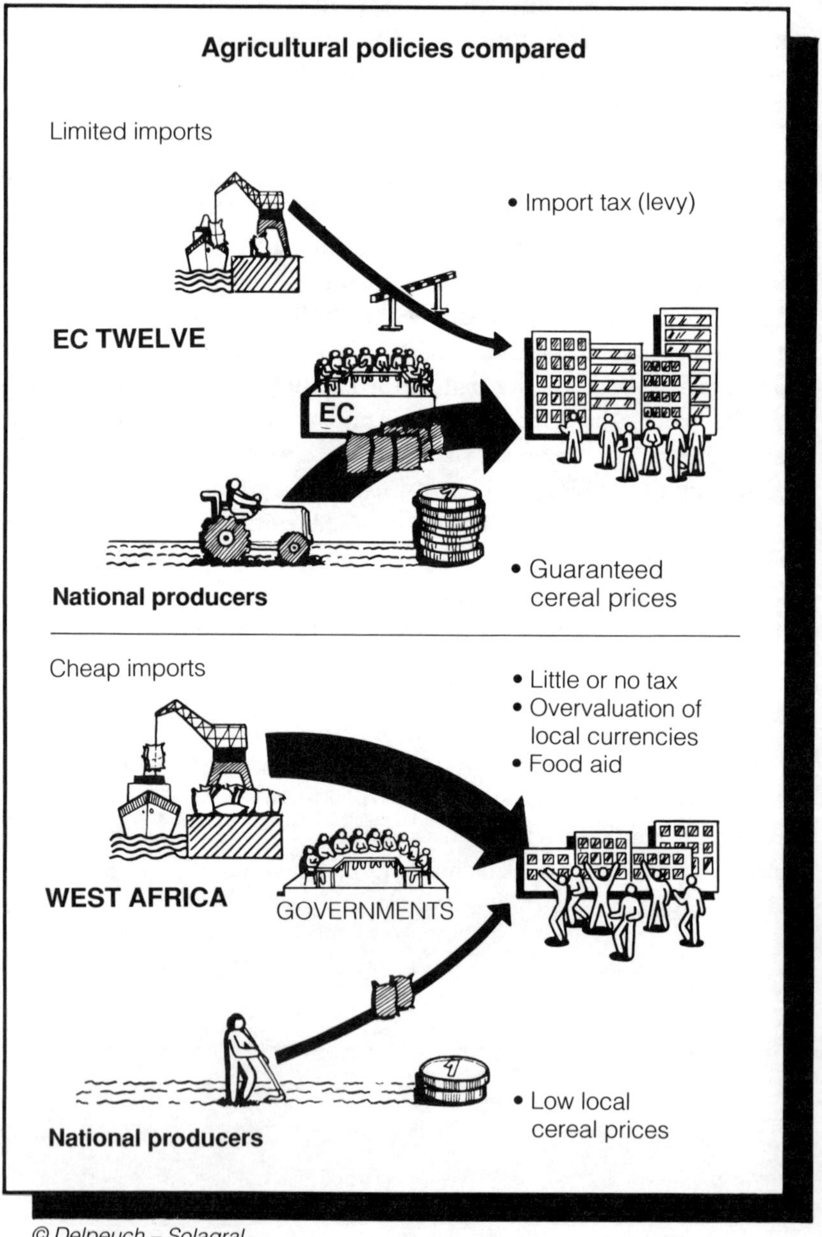

© Delpeuch – Solagral

From its inception, the EC decided to guarantee its farmers a minimum price for several crops, with a preference for purchasing within the Community, and with a system of import taxes. The reform of the CAP agreed in 1992 has not called these principles into question, even though guaranteed prices for cereals are set to fall much closer to the world market price.

African governments have chosen for many years to do the opposite, facilitating cheap food imports in order to ensure food supplies to the cities and thereby avoid urban social unrest which might threaten their power. Without the incentive of good prices, African farmers cannot withstand competition from the world market because of:

- Few or no import taxes;
- Their country's recourse to structural food aid;
- Overvaluation of local currencies which encourages purchases abroad.

Despite the transport costs, Thai rice arrives in Niamey, the capital of Niger, cheaper than local rice.[1]

The promotion of cheap imports has thus discouraged investment in African agriculture and favoured a long-term shift in dietary habits towards imported cereals. When the Senegalese government decided to revitalise national millet production by giving the farmer a better price, the harvest was significantly higher... but could not be sold in Dakar, whose inhabitants had gradually stopped eating millet.

Aware that the low incomes of their farmers, who form the majority of the population, are an obstacle to the development of their countries, several African governments have signalled their intention to protect local cereal production from cheap imports. They are indirectly helped in this by the international financial institutions which, by insisting on debt repayment and the devaluation of the local currency, in practice limit the country's ability to buy imported food.

- *In the case of French-speaking West Africa, it was announced in January 1994, following a meeting between French officials, the managing director of the International Monetary Fund and the 14-member African Financial Community (CFA), that the CFA franc (CFAfr) would be devalued 50 per cent in relation to the French franc (FFr).*[2]

Foreign debt repayments reduce the food security of the poorest

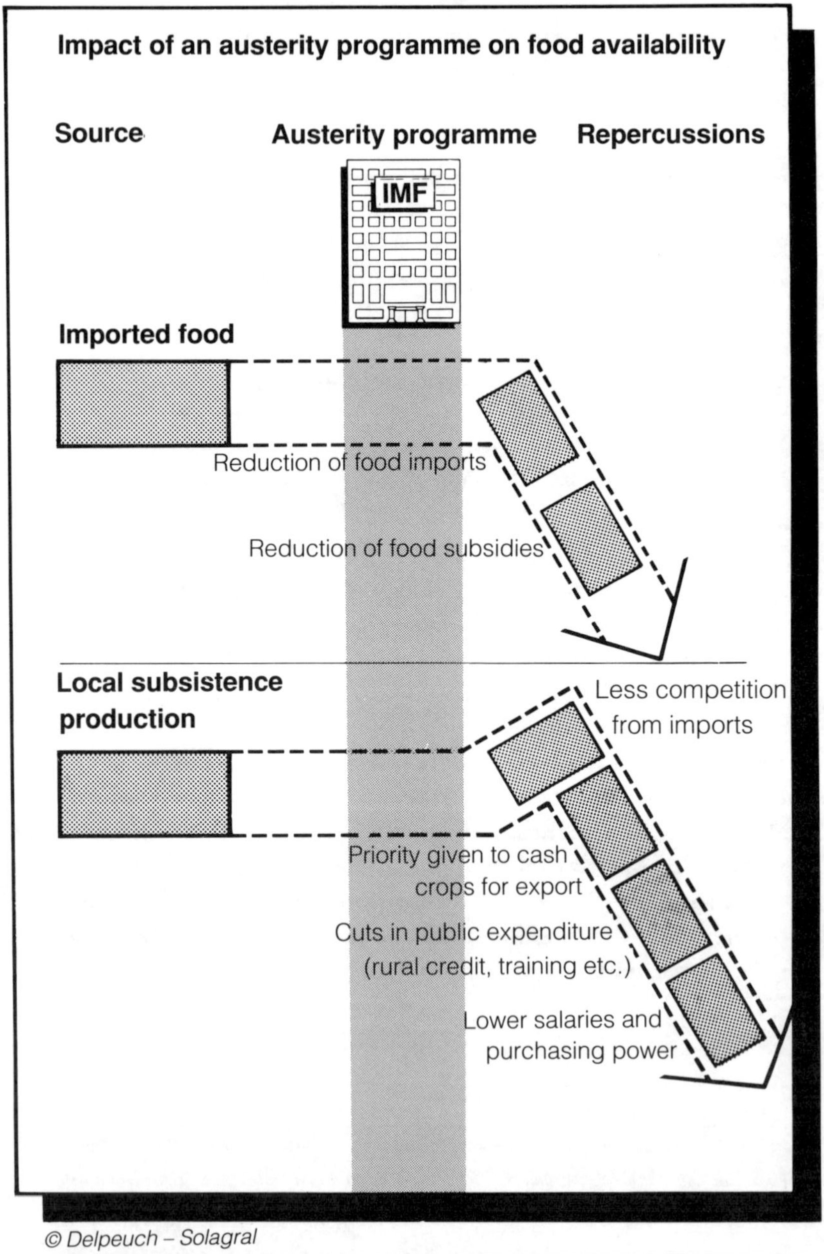

© Delpeuch – Solagral

The International Monetary Fund (IMF), before agreeing to a Third World country rescheduling its foreign debt, usually demands in return a package of measures intended to:

- Increase foreign exchange earnings by favouring cash crops aimed at the world market; soya, sugar cane and chicken are thus given priority in the allocation of fertiliser, credit or guaranteed prices, to the detriment of subsistence crops like beans, rice or cassava;
- Save foreign exchange by cutting public expenditure (health, education, food subsidies) and devaluing the national currency; imports, as they become more expensive, are held back whilst exports become more competitive.[1]

In fulfilling its financial mission, the IMF promotes economic measures which lead to a reduction in food imports. But it does not necessarily create favourable conditions for an increase in local subsistence food production, in spite of the rise in prices paid to producers. Structural adjustment programmes encourage, in the name of 'comparative advantage', specialisation by each country; cash crops are seen as the best way of generating additional revenue.[2] Hence the question posed by Julius Nyerere, the former president of Tanzania: 'Do we really have to starve our people to death to repay our debts?'

- *Increasing the volume of export crops does not automatically bring in more foreign exchange for a country; all depends on the price of its exports in relation to that of the manufactured goods which it imports.*
- *The military expenditure of developing countries has in turn been called into question by the IMF in the early 1990s; but cutting this expenditure is a difficult task when the more arms a government buys, the more aid it receives from Western countries.*[3]

Developing countries' food imports tend to increase with urbanisation

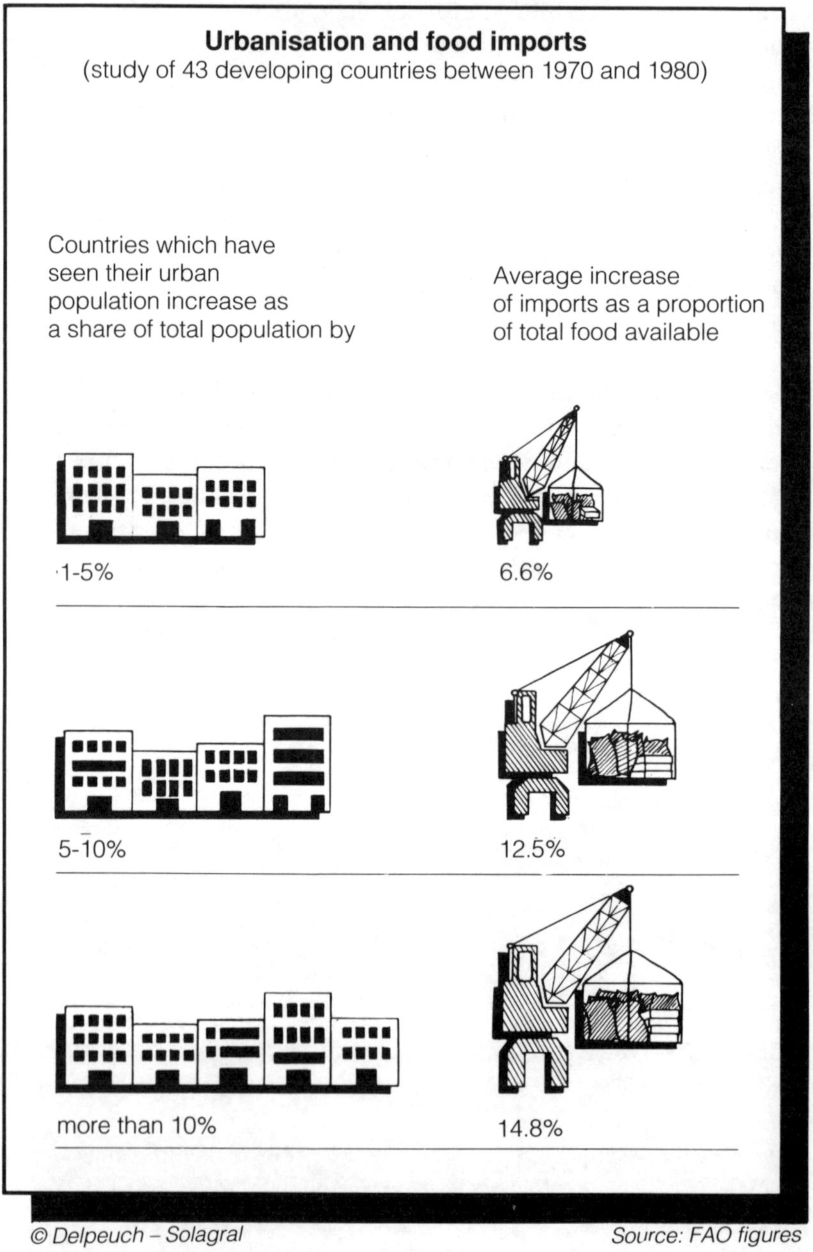

© Delpeuch – Solagral

Source: FAO figures

The proportion of the population living in urban areas depended for a long time on the size of the surplus in food production which farmers could supply.[1] In traditional societies, the production of food and agricultural raw materials required 70-75 per cent of the total labour force. The proportion of non-farmers, mainly urban dwellers, could therefore not exceed 25 per cent of the population. Infant mortality in towns and cities was higher than in the countryside. Starting in England in the 18th century, the agricultural revolution was to make surpluses – and therefore rising urbanisation – possible. Urbanisation then spread worldwide with progress in food transport.

Given that the major cities of the Third World are well linked to ports and railways, their inhabitants have become used to consuming imported cereals (wheat and rice), rather than traditional local foods (tubers, millet, sorghum). According to the United Nations' Food and Agricultural Organisation (FAO), there is a clear correlation in developing countries between the growth of urbanisation and the growth of food imports.[2] Does this mean that people are better fed in cities than in rural areas? It would seem so for the average urban family. But in the shantytowns, energy requirements seem harder to satisfy than in the rural environment. However, with the help of nearby health services, children from the shantytowns often enjoy a better nutritional standard, measured by height and weight, than in rural areas.[3]

- *It is in Africa that the future urban population explosion will be the greatest. While only one city had more than a million inhabitants in 1950, there were 19 in 1980, and there will probably be more than 60 in the year 2000. In Senegal, 10 per cent of rural children, but 'only' 5 per cent in the cities, have a weight 20 per cent below the norm in terms of their height and weight.*[4]
- *In Nigeria food imports have continued to increase even when the urban population has started to fall. This special case does not negate the general trend, with urbanisation, towards rising food imports.*

Cash crops often use productive resources at the expense of subsistence crops

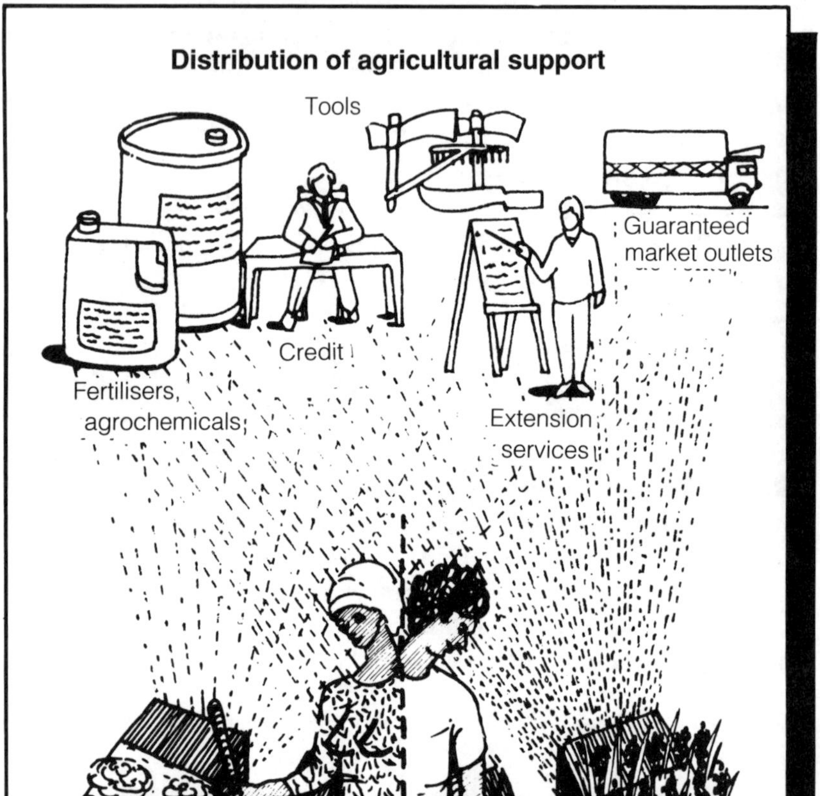

© Delpeuch – Solagral

To the detriment of subsistence crops, export crops were given priority during colonial times, and this has continued to be the case since independence. Export crops are often the only way for a country, particularly in Africa, to earn the foreign exchange needed to buy manufactured goods… and food.[1]

Locally, export crops can stimulate subsistence production as a result of the introduction of new techniques (animal draught, integrated crop and livestock farming). But they also often act as a brake on its development. Cash crops are given priority in the allocation of inputs (fertilisers, agro-chemicals), and in terms of research, extension services and credit, and guaranteed prices and markets. This unequal allocation of inputs is even more marked on plantations – a 'modern' agricultural system which makes little use of labour. The two systems compete for available working time, for labour and even for cultivable land. Areas planted with black beans, the staple food of the Brazilian poor, have shrunk as the more profitable soya and coffee have developed.

Does a country have an interest in reserving some of its land for export crops, using the income generated to buy imported food? The answer will vary, depending on the year, the country and the state of the world commodity markets. For many developing countries, agriculture will be the most likely sector to create employment in the future. Subsistence farming rather than export crops, for which world demand is stagnating, should be given priority in order to ensure at least a minimum degree of food security.[2]

- *Cotton in West Africa has an undeniable advantage over subsistence crops: it provides a guaranteed income and the farmer is paid cash.*[3]
- *Plantation monoculture has, from the beginning, excluded subsistence crops. In Brazil, slaves in the sugar cane plantations were fed imported cod from Newfoundland and dried meat from the Sertão; in the Caribbean, they ate salted beef and wheat from North America.*[4]

Food aid often discourages local production

Food aid: criticisms (•) and proposed reforms (^)

Frequent recourse to food aid:

GOVERNMENT

- • Prevents the strengthening of local agriculture
- ^ The revenue from food aid sales should be allocated to the development of local agriculture ('counterpart funding')

PRODUCER

- • Competes with local production
- ^ Food aid should be replaced by its financial equivalent in fertiliser and support for production and storage ('substitution actions')

CONSUMER

- • Leads to consumers acquiring a taste for foods which are little produced locally
- ^ Food aid should be bought, if possible, on regional markets, in the form of local staples ('triangular aid')

DONOR

- • Is determined by the size of European stocks
- ^ Food aid should be limited to emergency use, and to programmes aimed at making recipients self-sufficient

© Delpeuch – Solagral *Source: Solagral*

There are situations where emergency food aid is essential: drought, floods or a sudden influx of refugees. Yet emergency aid represents less than 10 per cent of total food aid.[1] The United States, the sole supplier until into the 1960s, and now the EC, are currently the two major donors. Almost systematically relied upon by some governments, food aid in non-emergency situations has been criticised on numerous counts:[2]

- It does not always benefit those who most need it (this is a criticism of distribution systems rather than of aid per se);
- Although a practical means of remedying food deficits, it does not provide an incentive for recipient governments to encourage their own agriculture;
- It reinforces demand for imported products; millet, sorghum and local maize today account for less than 10 per cent of food intake in Dakar;
- It promotes amongst the recipients a mentality of dependency on outside help.

These criticisms have led to a reformulation of food aid policies. Canada has cancelled maize shipments and converted them into credits for the purchase of local millet and sorghum in Mali and Niger.[3] The EC, which has transferred food aid management from the Agriculture Directorate (in charge of the Community's surpluses) to the Development Cooperation Directorate, is now encouraging substitution purchasing. Are donor countries therefore ready to give priority to local purchases rather than to the principle of 'shifting' their own surpluses? Are they ready to reduce their subsidised exports onto the world market, which act as a disincentive to production in Third World importing countries?

- *In 1992, following excellent harvests in the previous season, donor commitments of coarse grains were far higher than requirements in several African countries (Burkina Faso, Mali, Niger and Chad)*

The use of cereals for animal feed is gradually spreading in developing countries

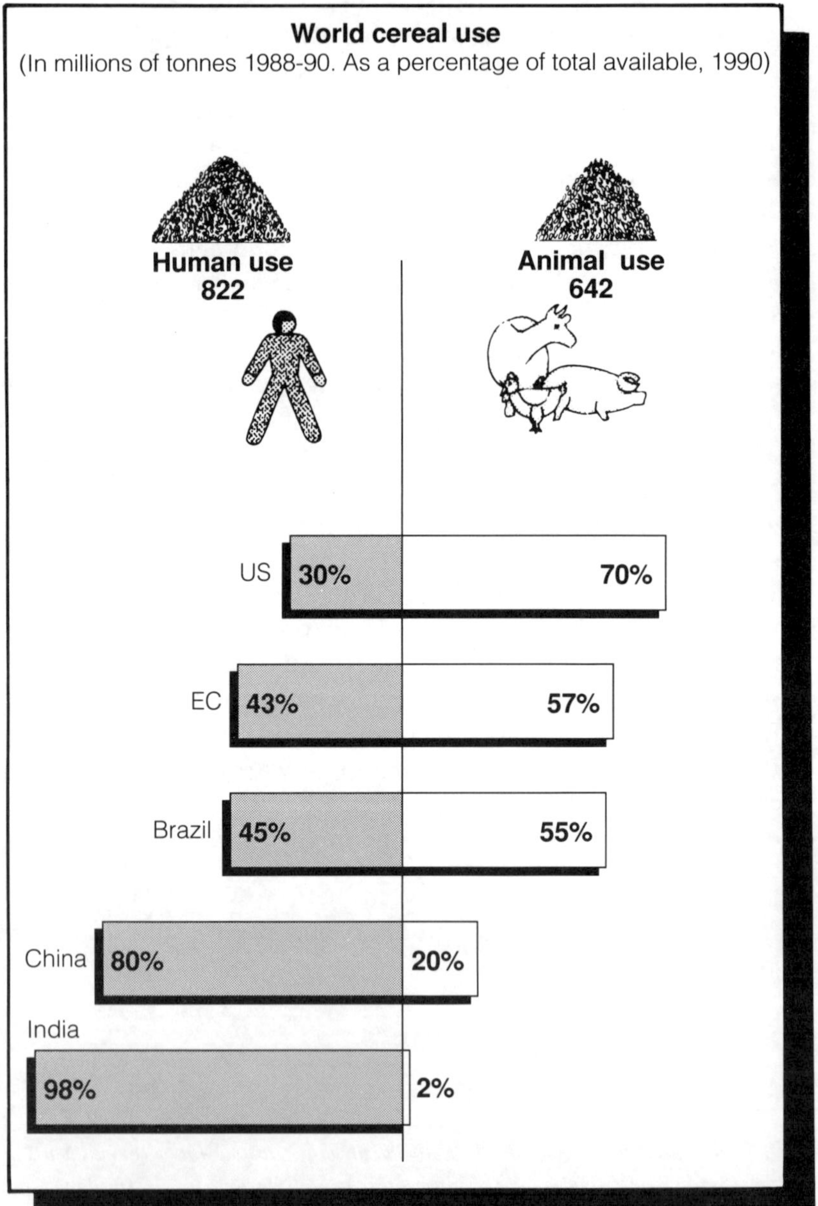

© Delpeuch – Solagral

Source: FAO

Nowadays, nearly half of the world's grain is consumed by animals. In the industrialised countries two-thirds of cereals are used as cattle feed: wheat, maize and barley in Europe, mainly maize in the United States and Japan.[1] The developed countries are the main users of feed cereals. North America and the former Soviet Union absorb 50 per cent of total world consumption. Animals are also consuming more and more of the cheaper cereal substitutes: cassava, corn gluten feed, citrus pulp and so on.

In the Third World the vast majority of cereals are still used for direct human consumption. The proportion of cereals used for animals remains low. But it is increasing rapidly: 5 per cent of wheat in 1975, 20 per cent in 1990. Animal feed for the Third World has become the market of the future for cereal exporters.[2]

For the last 20 years, trade in feedstuffs has grown considerably faster than their production. It has been the driving force behind the intensification of agricultural trade. It is the rise in demand for animal feed in the industrialised countries which has shaped agriculture in countries like Brazil and Thailand. First conceived in the 1950s as a way of absorbing cereal surpluses, cereal-based feed concentrate and cakes are an easily transferable technology. Their spread has been encouraged by US trade associations (the US Feed Grains Council and the American Soybean Association) in Japan, in Europe, and then in the Third World (Taiwan and South Korea, for example).

- *The per capita ratio between the plant biomass harvested and the food consumed by humans is ten calories to one in France. This 'loss' can be accounted for by agricultural exports and agro-industrial waste, but above all by the poor productivity of converting grain into animal products. Five calories are required to produce one calorie of milk, 7-20 to produce one of meat (depending on whether it is poultry or beef).*

Industrial fishing is costly in capital and energy and is overexploiting coastal resources

Industrial and small-scale fishing in Senegal (1989)

Type of fishing	Industrial	Small-scale
Proportion of fish thrown back in the sea	45%	5%
Consumption of fuel per tonne landed	450 l	190 l
Total numbers employed	3 000	35 000
Total landed catch (1989)	28%	72%
Share of fishing-grounds (in tonnes)	60%	40 %
Budget allocation in national plan (1983-89)	77%	23%

© Delpeuch – Solagral

Source: J.Weber and A.Fontana

World harvests from the sea increased more than six-fold between 1950 (16 million tonnes) and 1991 (97 million). The share of the Third World went from a quarter to more than half the world catch. One-third of the world catch now goes onto the international market.

The industrialised countries, given that their own stocks have in many cases reached near-exhaustion, import more and more fish for human and animal consumption. Almost a third of the world catch, often species suitable for human consumption, is processed into fishmeal to feed animals.

The development of industrial fishing fleets has made it possible to exploit stocks which had hitherto been left untouched or underexploited. But it has led to problems:

- Although in theory intended for deep-sea fishing, in practice trawlers often end up competing with small-scale fisherfolk working in their traditional waters, showing no respect for areas (where these exist) reserved for small craft;
- Technically highly-efficient, some trawlers pillage coastal waters without concern for stock renewal;
- Controlled by foreigners, they fish mainly for export. An agreement signed under the Marcos dictatorship meant that 90 per cent of Japanese catches in Filipino waters were exported to Japan. The Philippines, meanwhile, imported tinned sardines and mackerel from... Japan;[1]
- Industrial fishing, because it requires high levels of investment in technology and creates little employment,[2] is not a model for Third World countries. International organisations now admit having encouraged it too much, to the detriment of small-scale fisherfolk who supply the bulk of fish consumed in developing countries.

- *Fish is the only source of animal protein for one billion Asians.*[3]

Who will benefit from biological progress?

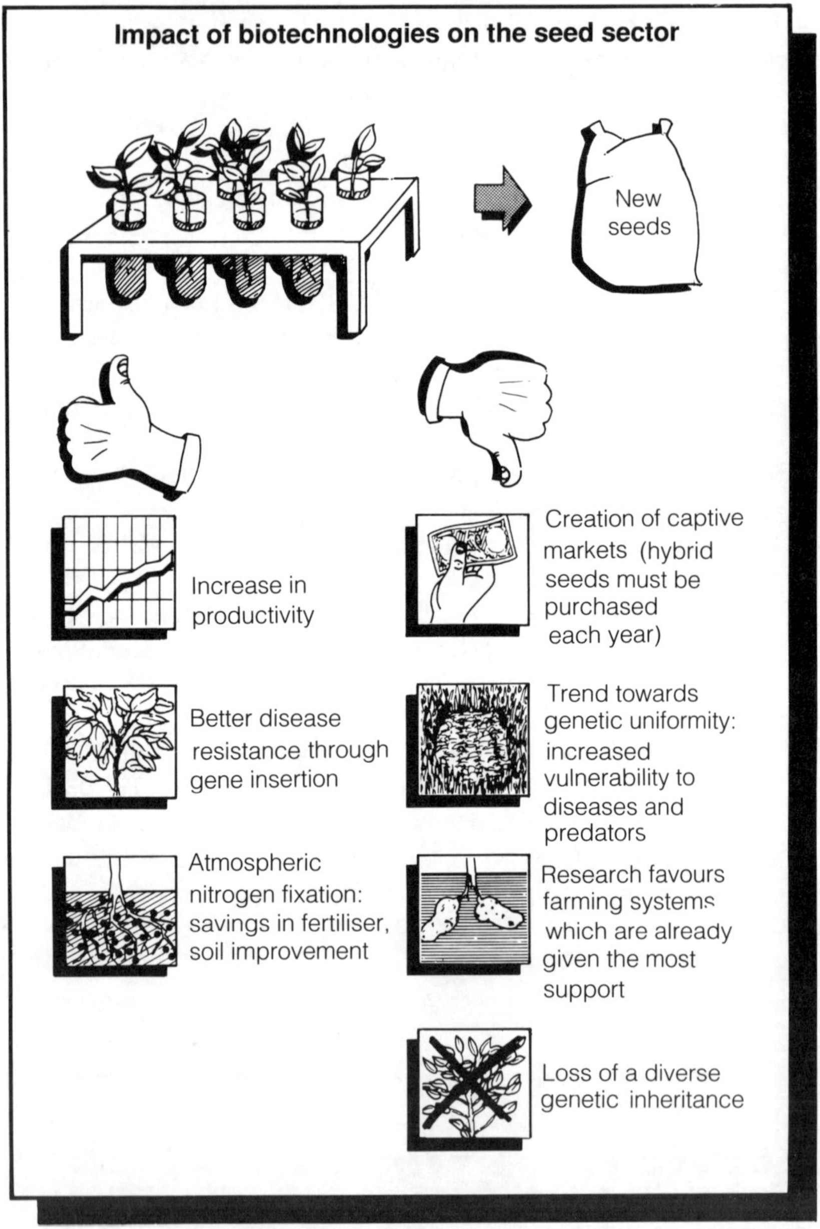

© Delpeuch – Solagral

Biotechnologies applied to seeds enable crop yields to be increased and production costs to be reduced. And when it is discovered how to transfer to cereals nitrogen-fixing properties already present in leguminous crops (beans, peas), substantial savings of nitrogen fertiliser will be made. Similarly, transferring resistance to a disease from a wild species to a cultivated variety will avoid costly pesticide treatments.

With higher yields and lower production costs, does biological progress offer hope for the Third World? Despite spectacular research, nothing is guaranteed.[1] The spread of a few rapidly multipliable varieties (test-tube multiplication makes it possible to produce 200,000 apple trees per year from just one bud)[2] will lead to a loss of genetic diversity. Crops become more vulnerable to the vagaries of the climate and diseases. Traditional varieties are abandoned, leading to the danger that their potential in terms of genetic diversity, as an irreplaceable reservoir from which new varieties can be drawn, will disappear. But above all, the scale of funds devoted to research leads to the creation of captive markets: in order to make a return on investment, each company seeks to sell as many seeds and related pesticides and fertilisers as possible. A company therefore channels its research into hybridisation, forcing the farmer to purchase seeds every year, and into varieties for which there is strong market potential, in other words, those often already grown in the most dynamic agricultural regions. Research into sorghum, sweet potato and *niébé* (an African bean) is at an early stage and little developed: these markets are judged not to be very profitable.[3]

- *Intended to conserve the world's genetic inheritance, gene banks store at low temperature millions of samples of cultivated local varieties and wild plants. Twenty-one of the 43 gene banks forming an international network are in developing countries.*[4]

Advertising alters consumer behaviour without concern for local priorities

© Delpeuch – Solagral

Source: WHO

Advertising is all the more effective in the Third World because there are few brand-names to attract the consumer and because consumer and public health protection is little developed.[1] When a multinational company selling alcoholic drinks decides to 'invest' in a country, that country's health and education budgets look remarkably small in comparison!

Industrially prepared drinks and foodstuffs such as milk concentrate, baby foods, soft drinks, tinned fish and tomatoes enjoy enormous prestige in developing countries. They are easy to store and easy to prepare. But they are also expensive, and often inappropriate for local nutritional needs. Each Mexican drinks an average 65 litres of soft drinks a year, three times more than milk, which is cheaper and more nutritious.[2] With its advertising budgets, the food industry can impose products of little nutritional value. Stock cubes, which are gaining in popularity in Africa, are easy to prepare but do not have the nutritional qualities of the traditional sauces they are replacing. Cigarettes, despite their danger to health, are presented as a symbol of social success; their consumption is increasing faster than population growth in every region of the Third World.

- *More cigarettes are smoked per capita in the developed countries (2,000-3,500 per year) than in the developing countries (less than 2,000, not counting roll-ups which are very popular). But while tobacco consumption is falling by 1 per cent annually in industrialised countries, it is increasing by 2 per cent in the developing countries.*[3]
- *Faced with increasing restrictions on advertising in the North, tobacco companies are looking for new markets in the South, where several have been selling cigarettes withdrawn from sale in developed countries for having too high a tar and nicotine content.*

Environmental degradation often has social causes

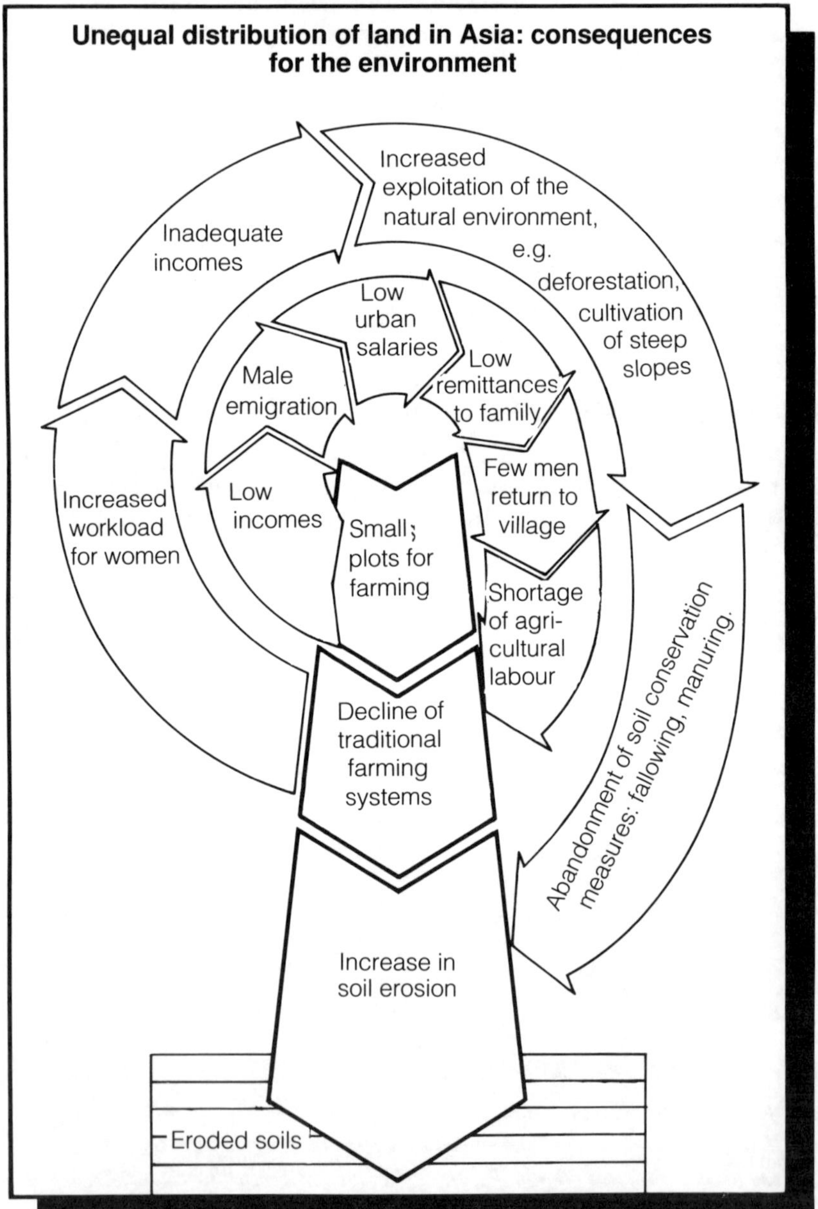

© Delpeuch – Solagral

Source: P. O'Keefe

Poor Third World populations know their natural environment well. They depend on it for their survival: food, fuel (wood and dung), animal fodder, medicinal plants, posts and roofing material for building.[1] But population growth means that the need for more land to farm, and for firewood, takes priority over long-term ecological considerations.[2] In the absence of alternatives, the poor are forced to degrade their environment. Unequal land distribution provides many examples:

- In Kenya and India, the scarcity of cultivable land leads the men to migrate to the towns and cities in search of work. Left on their own, the women only have time to deal with the most urgent tasks, and traditional practices for maintaining soil fertility are abandoned.[3]
- In Haiti, the insecure status of tenant farmers provides little incentive for the maintenance of irrigation channels.[4]
- In the Andes, indigenous farmers driven out of the valleys by settlers have had to farm mountain slopes which are very susceptible to erosion.
- In Ethiopia, the Awash River basin was developed by foreign companies for growing cotton and sugar cane. This has led the less fertile soils, to which the herds of the Afar shepherds have been driven back, to become seriously overgrazed.[5]
- In Afghanistan, farmers driven away by war (burnt harvests, mining of tracks) have given up their land to join the refugee camps.

No amount of regulation will protect the environment while the poor have no choice but to exploit it for their survival. It was economic development which saved the Swiss forests, threatened with deforestation in the 19th century in the same way that Third World forests are threatened today.[6]

Forced collectivisation has failed in agriculture

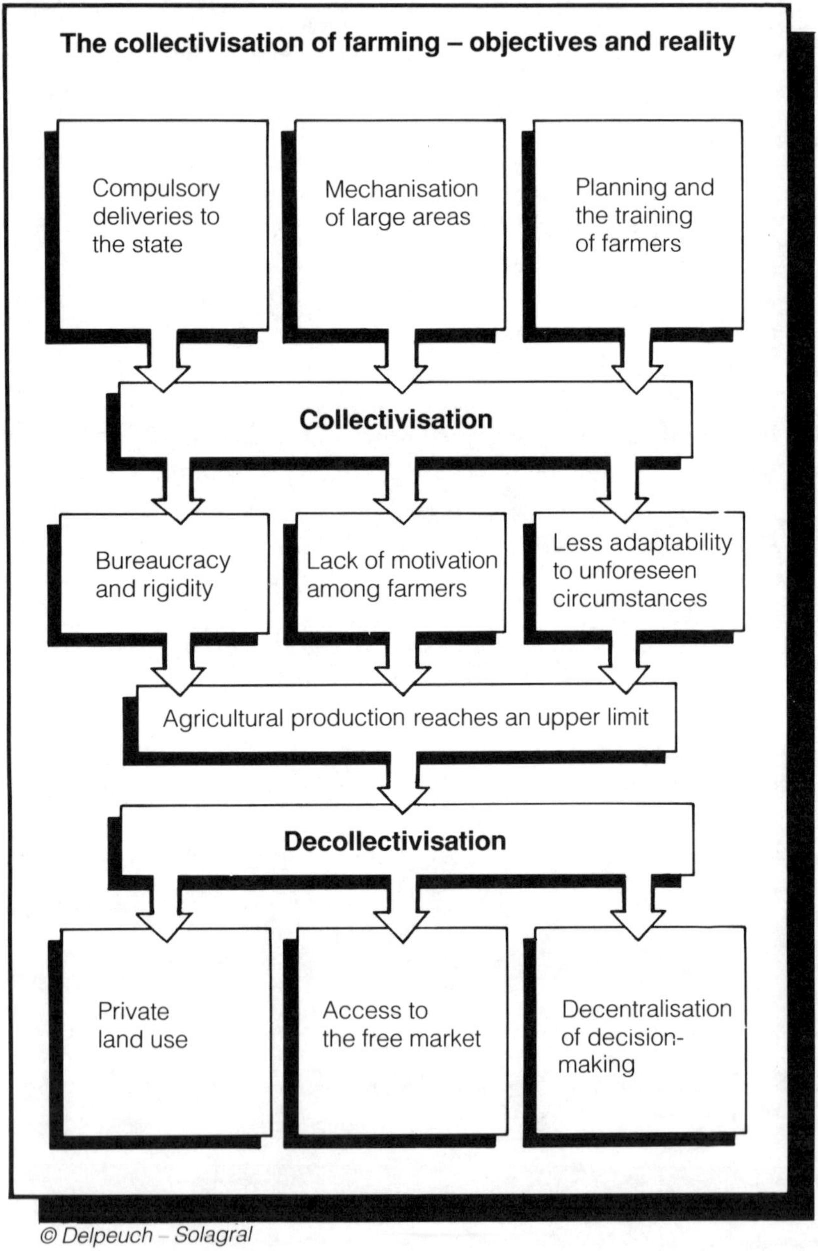

© Delpeuch – Solagral

The collectivisation of agriculture has often been intended by policy-makers to serve a dual purpose: increasing agricultural productivity and controlling farmers. The organisation of collective work was supposed to increase yields. Agricultural produce was handed over to the state, which took responsibility for its distribution to the population. This mode of production, tried in the Soviet Union from the 1920s, revealed its weaknesses as soon as farmers' membership of the cooperatives became compulsory.[1] Lacking motivation, farmers saw little benefit in a system which was often imposed by people who did not come from the village. Centralised planning quickly proved to be ill-suited to the diversity of situations found in farming like climate changes, soil varieties and cropping calendars.

At the time of decolonisation, several other countries opted to collectivise their agriculture: cooperatives in Vietnam, self-managed estates in Algeria, state farms in Cuba, Ujamaa villages in Tanzania. Although there may have been social gains with the establishment of collective services (irrigation systems, schools, roads and so on), the agricultural results have proved disappointing everywhere. Many of these countries are today attempting to return to production methods which give greater autonomy to farmers: private land use, access to the free market for the sale of produce and decentralisation of decision-making.[2] But, because of the lack of other local opportunities, Russian 'peasants' are reluctant to leave state farms.

This move towards decollectivisation does not solve all problems. There is a danger that it will widen income disparities between families and regions. Nevertheless, in China and Vietnam, it has led to new growth in agricultural production,[3] thereby confirming that a small family farm is often more efficient than a big farm.[4]

- *The forced collectivisation of Ukrainian agriculture caused the death of up to 7 million rural people during the great famine of 1932-33.*[5]
- *The success of the Israeli kibbutz demonstrates how effective collective farming can be when it is the result of free choice. But Palestinian farmers were excluded....*

3. ACTING ON THE FACTS

Food strategies and protectionism

Land reform – Diversification of models

Triangular aid, codes of conduct, agricultural policy: acting in the North

The voice of the South

Give Third World farmers a chance

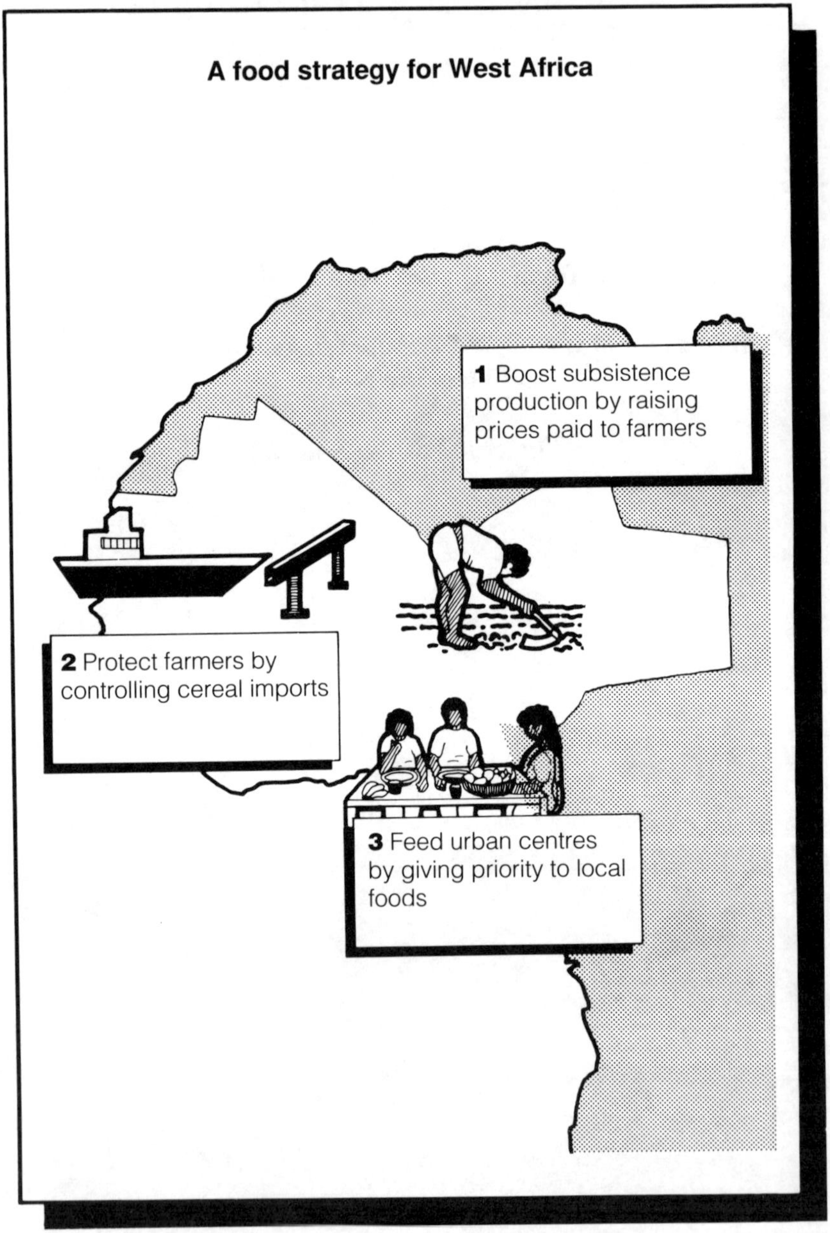

© Delpeuch – Solagral

Forced to compete on the world market with its cut-rate prices, Third World farmers have little chance of developing their subsistence production and of regaining urban markets. A Third World country's search for food security therefore requires a two-fold strategy: boosting local food production and setting up, with neighbouring countries if possible, a protected common agricultural market. Food imports would then be used solely as a means of regulating supply,[1] or even as insurance against bad harvests.

Several countries (including Mali, Zambia and Rwanda) have attempted, with EC support, to implement such a food security strategy involving the following:

- Support for production: rural credit, training and agricultural research, emphasising the use of local resources;
- The improvement of post-harvest storage, transport infrastructure and marketing networks targeting the cities;
- The processing of local foods such as millet, sorghum and cassava, to make them as attractive and easy to cook as imported wheat and rice.

If these measures, currently adopted by the EC in the rural development programmes it finances, are to show tangible results, they must be complemented by national policies which enable the urban poor to gain adequate employment and income to buy their food: bread-winning before bread-eating.[2] West African countries and those of the Andean Pact are studying the feasibility of setting up common agricultural policies.[3] Pro tected regional subsistence production would no doubt cost more than importing. But it could be financed in local currency, an advantage for often heavily-indebted countries. It would also enable jobs to be maintained in agriculture. Taiwan and South Korea used such protection measures as a means of supporting the early stages of industrialisation and, combined with land reform, this has been one of the key elements of their economic success.[4]

Carry out land reform to boost small farmer agriculture

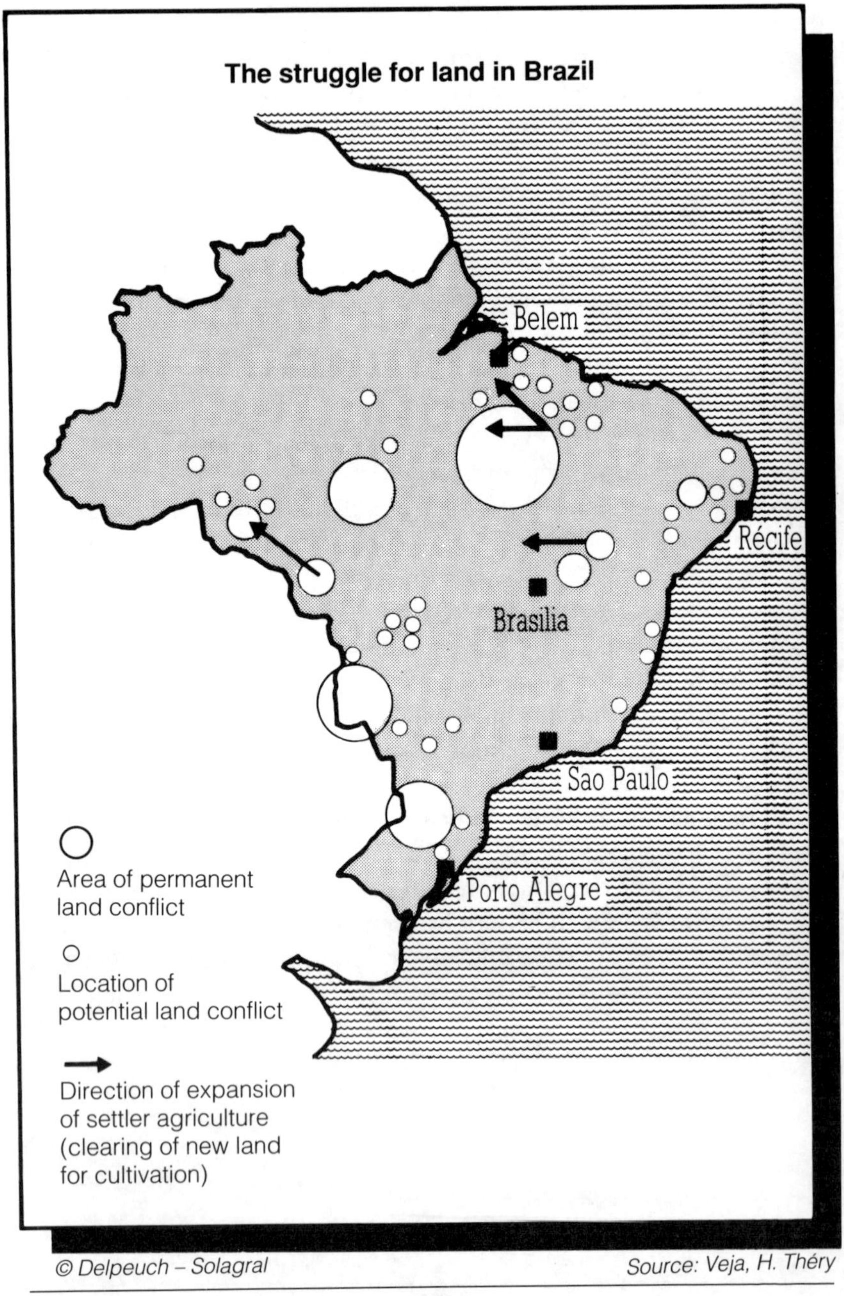

© Delpeuch – Solagral Source: Veja, H. Théry

The juxtaposition of huge estates and tiny farms is seen as the greatest injustice by poor farmers condemned to destitution as a result. Furthermore, large estates generally have a low level of productivity. Many studies in Latin America and Asia have shown that the average yield is higher on small farms. They are exploited more intensively and resourcefully than large estates, which are given over to extensive livestock and crops.[1]

Land reforms have been decreed in many countries to satisfy the demands of peasant farmers. But they have often been only partially implemented because of pressure from large landowners. Reactivating these land reforms and distributing land would give work to poor farmers and would increase the amount of food available. An essential complement to any land reform is a revision of rural tenancy laws, which should forbid excessive rents and guarantee farmers security of work on the land for several years. But access to land, in itself, is not an absolute guarantee of income for a family. There should also be fairer access to the means of production (water, equipment), to credit and extension services, to storage facilities and transport.

Third World countries are in debt and farmers' children are migrating to the cities. By redistributing the land, agrarian reform would allow small-scale farming to develop, which would increase the productivity of soil by absorbing the maximum labour with a minimum of capital.[2]

- *Brazil is a typical example of a country with enormous land resources. The large landowners have opposed agrarian reform by force. Since 1984 they have had murdered on average one supporter of land reform (a peasant or land lawyer) each day.*
- *Land reform is also a current issue in Europe. In Andalucía, in Spain, 400,000 farmworkers have no work for most of the year, yet 0.5 per cent of landowners own 54 per cent of the land.*[3]

Diversify agricultural development models

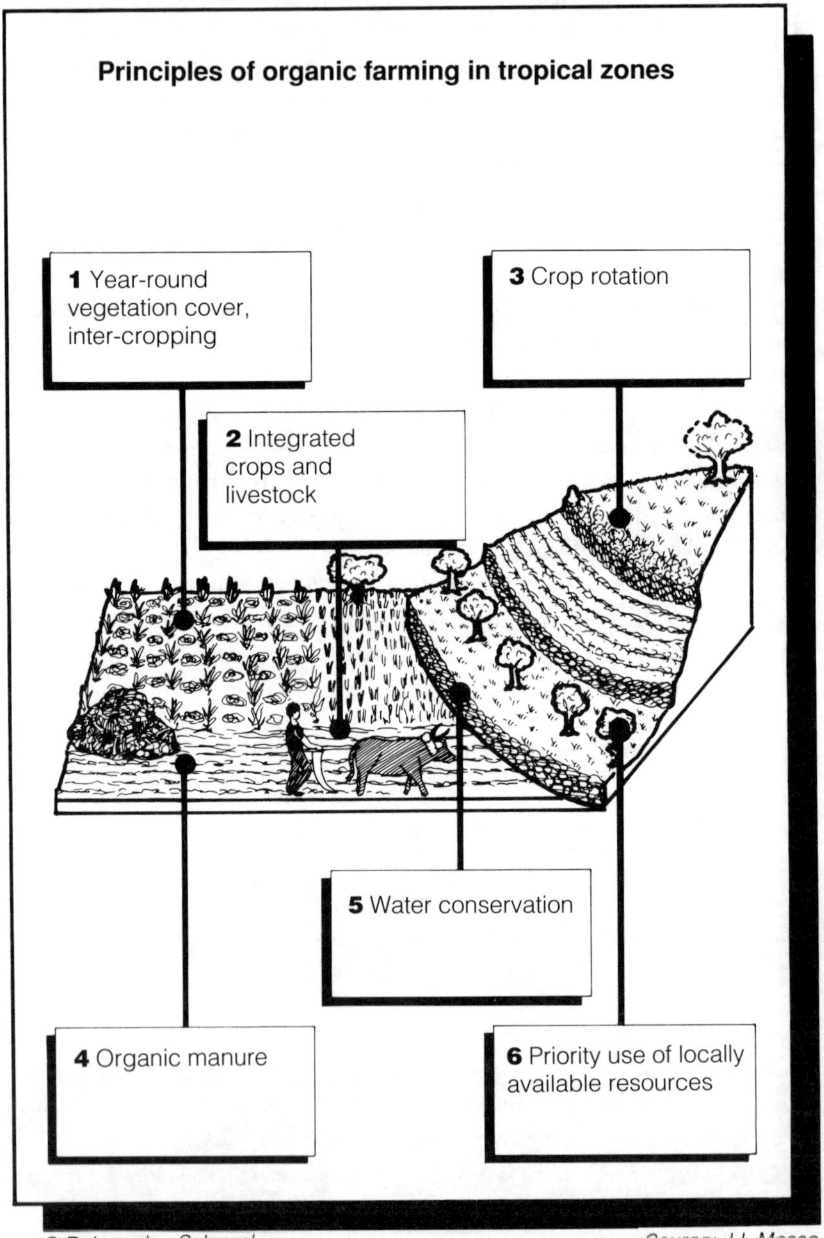

© Delpeuch – Solagral *Source: J-L.Messe*

Today, agricultural research and development involves criteria other than yield alone. The objective is no longer just to maximise production in absolute terms, but also to take natural and economic constraints into account. The limits of Europe's over-intensive agriculture are now becoming more apparent with the pollution of soils and water, factory farming on a massive scale, compulsory set-aside and so on. This crisis is encouraging research into other models of agriculture which reduce the use of external inputs: nitrogen fertiliser limitation; replacement of imported soya by local crops such as white clover,[1] sunflower and peas; giving priority to food quality; and organic agricultural methods.

In Third World countries, agricultural research is now rectifying the imbalances of the past. Rain-fed agriculture, although representing 80 per cent of the world's cultivated land area, has been much less studied than irrigated agriculture. Researchers are now working on tubers, legumes and traditional cereals which have been neglected for years in favour of just rice and maize. Inter-cropping and integrated crop-livestock-aquaculture systems, widespread in Asia, are regaining their status. The importance of market gardening in Sahelian Africa is beginning to be recognised.

Practiced in tropical zones, organic farming offers the opportunity to rely first and foremost on locally available resources: compost-based manure rather than imported fertiliser, animal draft rather than heavy mechanisation, biological pest control rather than purchases of chemicals for crop protection. Yields can be increased sustainably without either farmers or developing countries incurring debts.[2] These new trends in research will not change the agricultural landscape of the planet by themselves. But at least they provide the basic knowledge on which a different model of agricultural development, one that is more self-reliant and less wasteful, can be built.

Two ways of rebuilding Haiti's pig population

1 Intensive rearing: costly imports (feed, buildings)

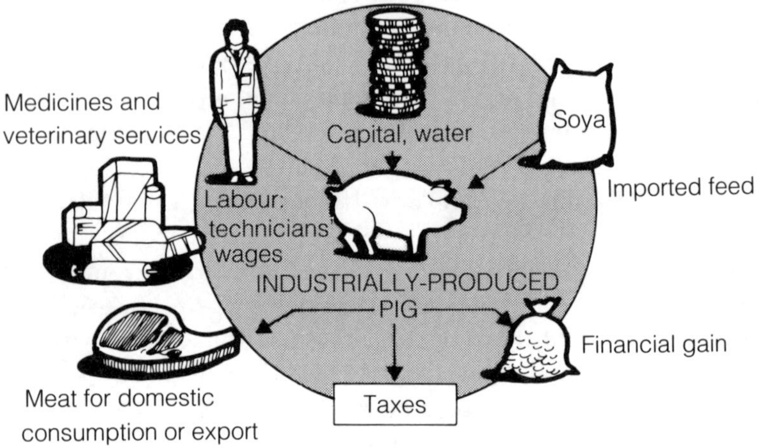

2 Traditional small-scale pig rearing: reduced costs of production

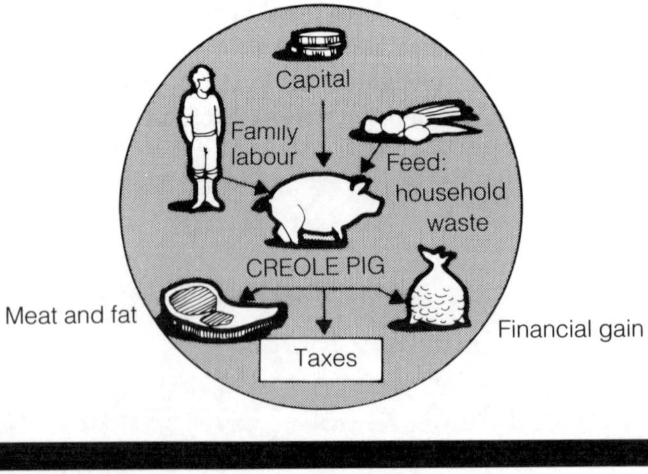

© Delpeuch – Solagral

There are often two technological choices for the same development project:

- The 'modern' approach, which is efficient in the context in which it was conceived, in other words, in an industrialised country;
- The 'appropriate' approach, which seeks to take into account what is locally available, in other words, for a developing country, plenty of labour but little capital.

The case of rebuilding Haiti's pig population illustrates this choice well. All Haitian pigs were destroyed between 1981 and 1983 following an epidemic.[1] What model of production should be reintroduced?

- Intensive rearing as encouraged by the United States? Improved pig breeds produce a lot of meat but require a highly organised chain of production. In Haiti, this would mean importing feedstuffs, buildings and specialists, all of which excludes poor farmers.
- The peasant approach, supported by non-governmental organisations (NGOs), which favours indigenous breeds? These pigs produce less meat but, since they are fed on agricultural waste and are more resistant, their rearing is virtually free and can therefore be afforded by all farmers. Profit per pig is higher.

In Mexico, some indigenous peasants select future maize seeds directly from their fields. Just as efficient as the hybrids sold by companies, the seeds selected in this way have the advantage of being free for peasants, as well as the guarantee of being better adapted to the environment.[2] 'Why impose "cut-price" technologies on the Third World?', some will ask indignantly. The answer is that people feel greater concern for a production system which they can make their own and maintain and improve themselves so that it is better suited to their needs.[3]

- *In Africa locally prepared baby cereal made of groundnut paste, flour, dried leaves and fresh milk can be as nutritious as imported milk-based preparations.*[4] *Algeria has therefore introduced a local flour made from wheat and chickpeas.*

Take the role of women into account

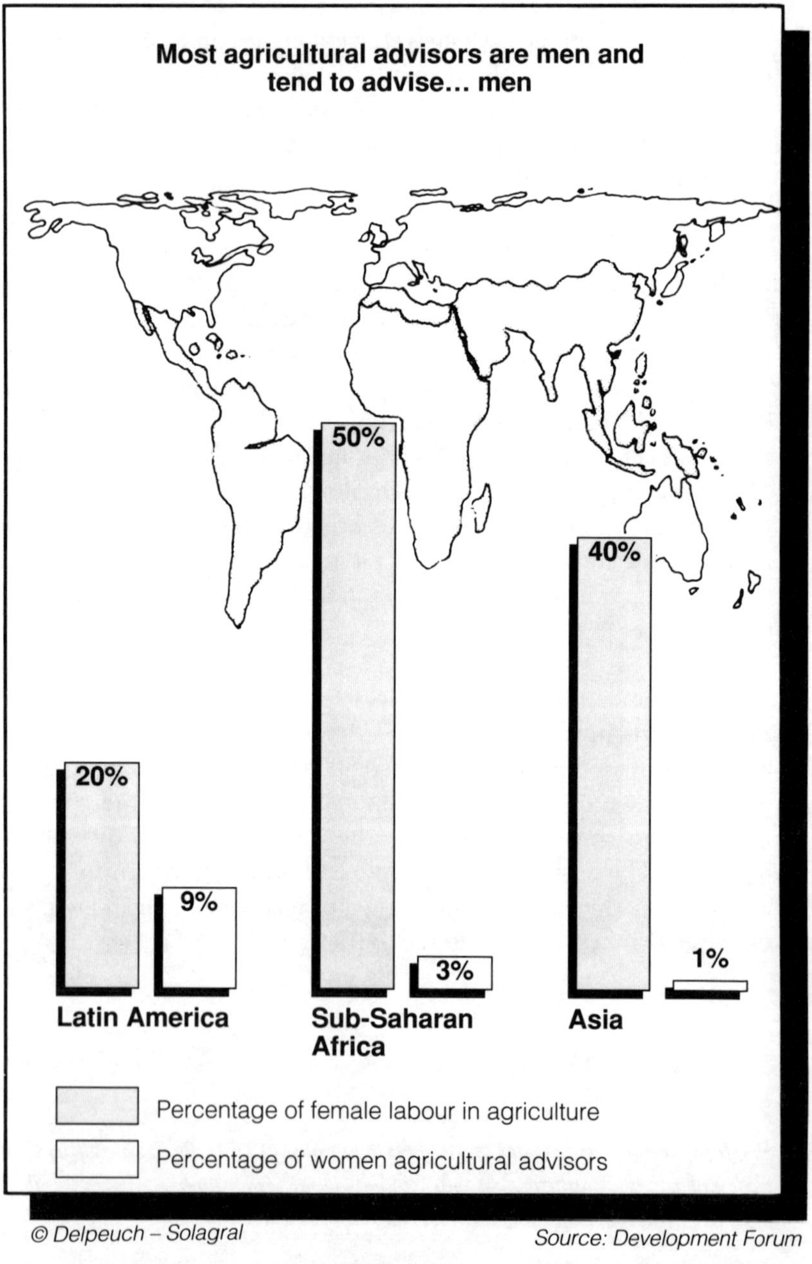

© Delpeuch – Solagral

Source: Development Forum

Many development projects ignore the role of women in food production:

- When a land reform is carried out or when an area becomes irrigated, women are frequently denied land of their own, even when their experience would justify land being granted (as with rice cultivation in Sahel);
- Agricultural mechanisation and the establishment of new marketing networks do not consider the jobs – traditionally women's jobs – that are lost in the process;
- Forestry policies neglect the collection – vital for poor women – of wood, kindling and fodder (up to five or six hours a day in India).[1]

Greater recognition of the role of women would be in keeping not only with demands for equality, but also with concern for the efficiency of food policies.[2] Among other things, this would involve:

- Women's access to credit, fertiliser, seeds and property deeds (in the case of individually granted titles) on the same terms as men;
- Women's representation in peasant and farmers' organisations and in agricultural support bodies;
- The development of agricultural research in vegetable gardening and inter-cropping, taking into account women's practices (appreciating the fuel value of waste or the nutritional value of leaves, for example);
- The improvement of small-scale processing techniques for cereals, oils, and fish, which are often the responsibility of women.

- *Rural development plans conceived by men often give priority to irrigation, but rarely to the provision of drinking water. 'If men were responsible for fetching drinking water, 230,000 Indian villages would not still be without it after 30 years of development planning,' remarked one senior Indian civil servant.*[3]

Opt for triangular food aid

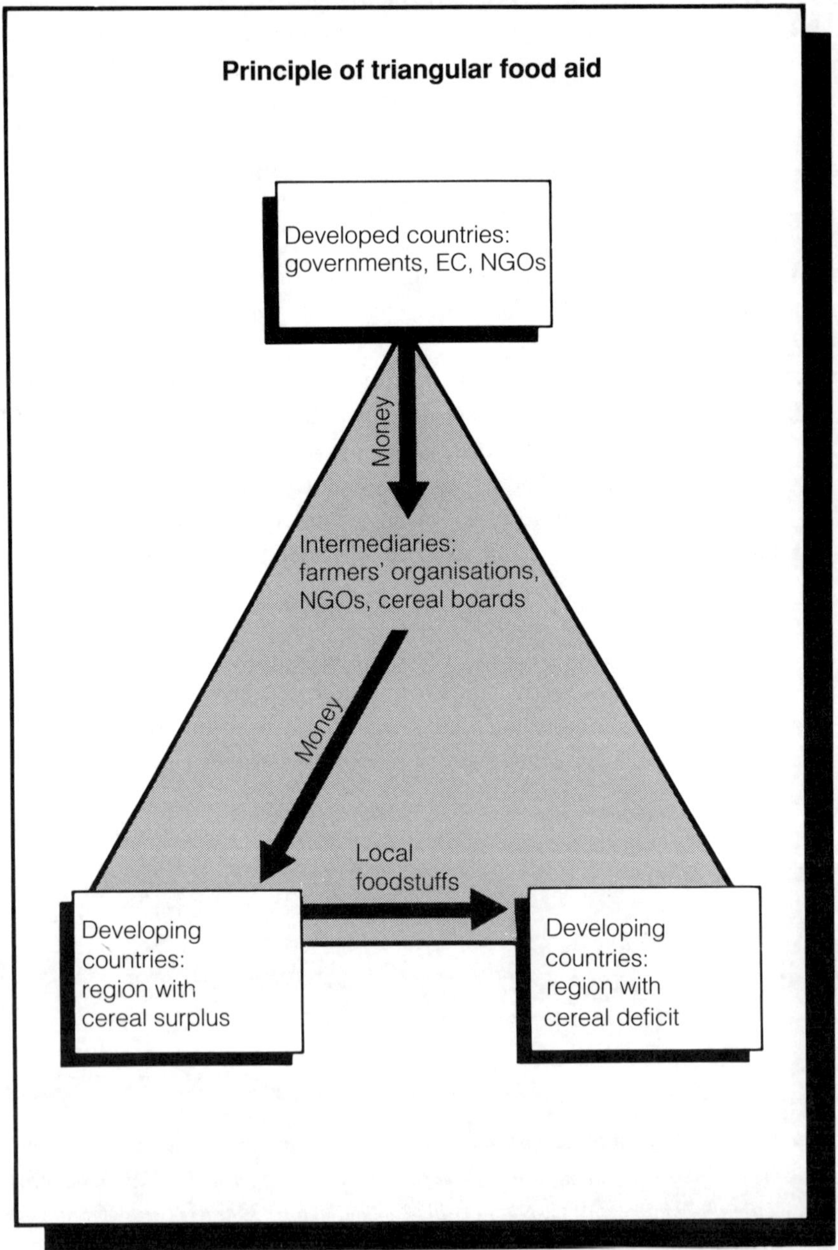

© Delpeuch – Solagral

Source: Afrique Verte

The concept of triangular food aid is based on recognition of the fact that while some regions of the Third World face food shortages, others – not necessarily very far away – may have a surplus available at the same time. Why not therefore replace shipments of cereals from the North with a sum of money to purchase local cereals from a surplus area for redistribution in a deficit area? Farmers' organisations or NGOs can take charge of the operation.

Compared with classic food aid, sent by industrialised countries which draw on their agricultural surpluses, triangular food aid has several advantages.[1] It does not interfere with traditional food habits and it provides new markets for farmers in surplus areas. But it also faces the difficulties of any new initiative, such as untested administrative procedures and the lack of local road networks. Assessing from which surplus area to buy is also difficult. A farmer may be tempted to sell grain for cash in hand, even if it means depleting his family's food stock.[2] Nevertheless, the merit of triangular food aid is that it boosts local trade and helps promote the idea of organised and protected regional markets. The beneficiaries of the aid cease being the passive recipients of Northern surpluses.

Relieve the burden of external debt

Measures proposed to relieve developing countries of their debt

1980s

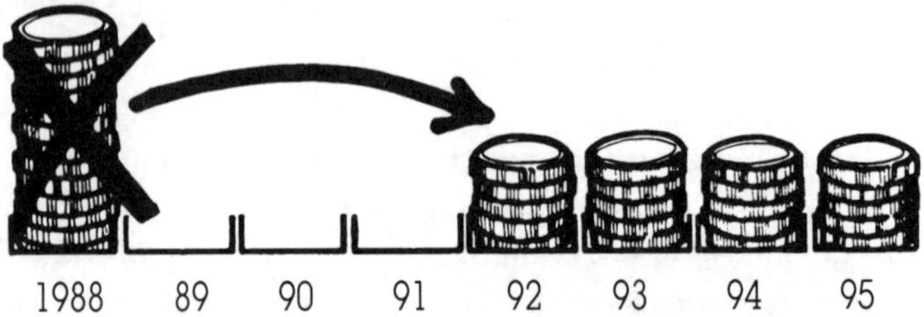

• Rescheduling with a moratorium of several years

1990s

• Cancellation of debts

• Debt-for-social projects or debt-for-nature swaps

© Delpeuch – Solagral

During the 1980s, the illusion persisted that debt rescheduling, coupled with the postponement of repayment deadlines and lower interest rates, would enable Third World countries to honour their debts. Their total foreign debt has continued to grow, however, from US$760 billion in 1982 to US$1,700 billion in 1992.

In 1989, the so-called Brady Plan officially accepted what was clear from the evidence: that developing countries would never be able to pay back the whole of their debt. Thereafter, the countries of the North adopted a new strategy: debt forgiveness, in particular cancellation, at least partial, of official debt.[1] The big commercial banks, for their part, had for several years been forced to accept a reduction in the value of the debts, traded at a large discount on the secondary market.

Encouraged by the World Bank, debt reduction programmes including cancellation ought to become the norm. But while these programmes make it possible (rather late in the day) to wipe the slate clean, they do not guarantee Third World states access to new, yet needed, financial resources.

- *Taking advantage of big discounts on the secondary market, some NGOs buy, in dollars, shares of the commercial debt of developing countries. In exchange, the developing countries receive finance, in local currency, for humanitarian or environmental protection projects. Although attractive, such social or environmental debt-swaps remain marginal – they comprise less than 1 per cent of the total – because of the lack of political will.*[2]
- *To finance the hosting of Guatemalan refugees in southern Mexico, in 1993 the United Nations High Commission for Refugees bought US$10 million worth of Mexican debt, an innovative operation allowing Mexico to repay its debt in pesos (thus saving foreign exchange) and to finance humanitarian programmes.*[3]

Encourage international codes of conduct

Main points of the WHO's code for the marketing of breast milk substitutes

1 Promotion of breast-feeding. Mention of its superiority in all documents aimed at mothers-to-be

2 Regulation of labelling on baby milk products; instructions for use in appropriate language; no deceptive illustrations

3 Ban on advertising of breast milk substitutes; no free samples to mothers; no free gifts or bonuses to health workers

4 Introduction of legislation in every country; monitoring of application of the code by WHO, NGOs etc.

© Delpeuch – Solagral

An international code of conduct is a list of recommendations drawn up under the aegis of an international organisation (WHO, FAO, Unctad etc.) to regulate commercial practices in a given sector. Such a code can be strictly voluntary, relying for its application on the consent of governments and companies, or can be intended for incorporation into the national legislation of the states party to the code.

Approved in 1981 by the World Health Organisation (WHO), the International Code of Marketing of Breastmilk Substitutes[1] was the fruit of work by NGOs which had denounced for several years, with the help of the media, the dangers associated with bottle-feeding in the developing countries. At the time, ten million babies and infants were suffering malnutrition and diarrhoea linked to unsuitable bottle-feeding. The Code, adopted in spite of the efforts of the companies to oppose it,[2] regulates their operations in order to protect children's health. Similar codes have been approved or are being discussed for pesticides, medicines, seeds and so on.

The adoption of any code is the result of a compromise between governments, the companies involved, which want the recommendations to be as weak as possible, and NGOs, which defend the principle of strict codes. The latter do not believe that total trust can be placed in the companies. Examples abound of dangerous products banned in developed countries (pesticides, certain brands of cigarettes) which companies have sold unscrupulously in developing countries. Monitoring the enforcement of an international code of conduct often poses problems. Governments drag their feet over the code's incorporation into national legislation, which is still the only way of giving it legal status. Resources for monitoring are inadequate. Many developing countries, for instance, have no procedure for controlling pesticides in their territory. Yet despite such weaknesses, international codes are useful safety-barriers. Their adoption, and the monitoring of their application, depend largely on the dynamic campaigning efforts of citizens' groups in the North and in the South.

Promote the participation of social actors in the South

The future of people in the Third World also depends on decisions taken in Brussels, New York or Geneva...

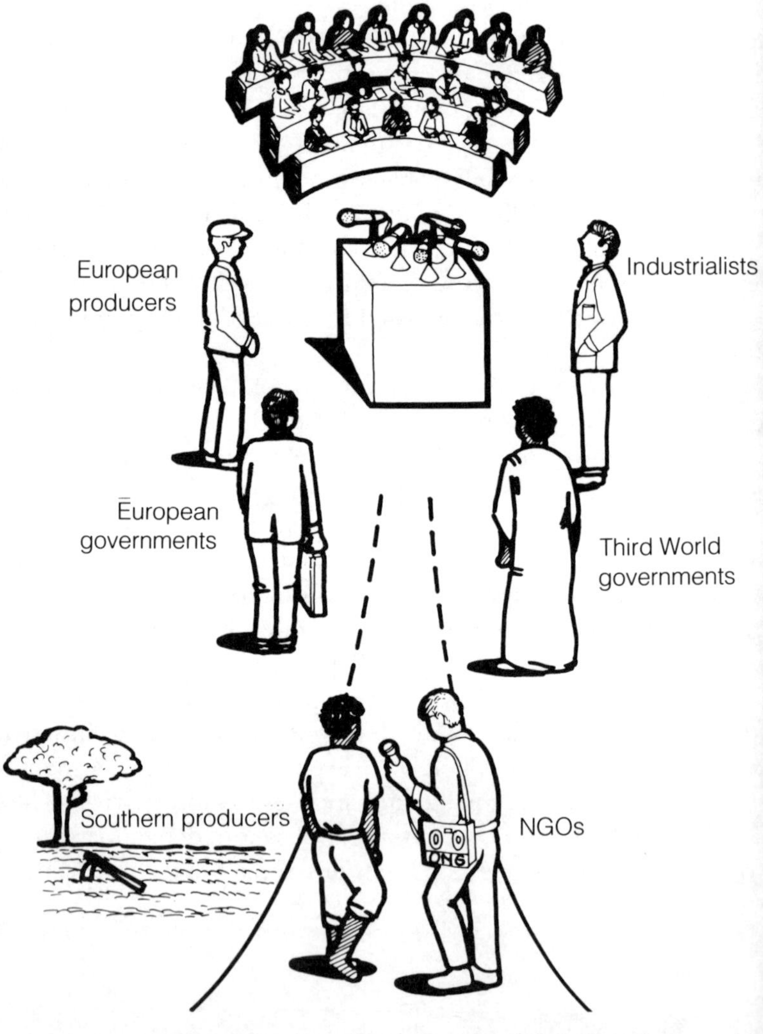

© Delpeuch – Solagral

In an increasingly internationalised economy, products travel more than the people who produce them. The fate of a Senegalese or Brazilian family partly depends on decisions taken by governments or Western companies. This is where NGOs can intervene. New forms of international cooperation must be devised, in order to complement 'conventional' development projects like funding a well in Mali. For example:

The exchange of information between social groups involved in the same production activities in the North and South. Brazilian and Thai farmers need to know how their soya or cassava are used in Europe, just as European farmers must debate the social impact of these imported crops.

The participation of social actors from both the North and South. At the World Fisheries Conference[1] in Rome in 1984, as during the renewal of the EC-Senegal fishing agreement in Brussels in 1992, representatives from Third World fishing communities invited by NGOs were able to make their voices heard.

Support for the efforts of social movements to organise. On every occasion where farmworkers have been able to organise, as in Kerala (India), salaries have tended to stay above the statutory minimum.[2]

Campaigns against injustice. North American pressure groups, which forced the Reagan administration to end support for the Marcos regime, made it easier for Filipino peasants to get their demands for land reform on the agenda. In the cases of Afghanistan, Chile or South Africa, the pressure of international public opinion has played a crucial role in giving democracy a chance.

Making international solidarity a factor in national policy- and decision-making. It is ridiculous to provide a certain percentage of GNP in support of Third World countries if, at the same time, 'by closing our markets or competing with their production, we make them poorer than they already are'.[3]

LIVERPOOL HOPE UNIVERSITY COLLEGE

Conclusion

World hunger is, like any other theme, subject to intellectual fashions. Agricultural collectivisation, the hope of the 1960s and then of the supporters of the Chinese model, was supposed to help save rural areas from famine. Its limitations have since become clear everywhere. Transnational companies and the West's share of responsibility, denounced in the 1970s, have since disappeared from the mainstream debate. Third World governments are now in the dock for having failed to prioritise agriculture, for having supported inefficient parastatal marketing bodies and for having enriched oversized bureaucracies. This book has attempted to retain what is valid in each of these critiques, whilst not forgetting the leading role of the economies and decision-makers of the North. What use is it talking to the inhabitants of Africa about the nutritional value of their traditional leaf-based sauces if at the same time a giant food company's sales of stock-cubes are booming?

Land reform may well be essential in Bangladesh and higher agricultural prices desirable in Africa, but it is not up to people in the North to take those decisions. It is, however, their responsibility to support the farmers' organisations which are making these demands; to ensure that Northern governments do not make such reforms impossible by their support for authoritarian regimes, by promoting inappropriate projects or by flooding the market with their surpluses; to remember that it is all too easy to advocate wholesale economic liberalisation for the South when the industrialised North protected its own agriculture until food security was achieved. It is also the role of people in the North to show the way by promoting agricultural policies which enhance equity between people and regions, and which respect the environment and rural communities.

Concern for the Third World will only cease to be a charitable afterthought in industrial societies when the North-South dimension of international relations becomes, among others, one of the premises on which social change is based. This inevitably means rethinking the role of public institutions and of political action.[1] The campaign network which succeeded through sheer perseverance, after 10 years of public lobbying, meetings and petitions, in securing the adoption of the International Code of Marketing of Breastmilk Substitutes has undoubtedly saved as many Third World lives as all emergency relief campaigns. Effectiveness in development work does not necessarily involve spectacular actions.

References

Page 4

1. Rudolph H. Strahm, *Pourquoi sont-ils si pauvres?*, A la Baconnière – Déclaration de Berne, 1986, p.219.

Page 5

1. Henri Rouillé d'Orfeuil, *Le Tiers-monde*, La Découverte, 1987. Yves Lacoste, 'Unité et diversité du tiers-monde' in *L'état du tiers-monde*, La Découverte, 1987, pp.7-22.
2. Domitila, *Si on me donne la parole*, Maspéro, 1978, p.254; Elisabeth Burgos, *Moi, Rigoberta Menchú*, Gallimard, 1983, p.330; Renée Colin-Nogues, Yakaré, Maspéro, 1982, p.240; Robert Linhart, *Le Sucre et la Faim*, Minuit, 1980, p.95.
3. Solagral, 11 Passage Penel, F-75018 Paris, France.

Page 6

1. Johnny Egg and John Igue, 'Echanges invisibles et clandestins', in *Intertropiques* No.22, May 1987, p.103.
2. Georges Courade, 'Une "révolution verte" pour l'Afrique?' in *Politique africaine* No.26, June 1987, p.103.

Page 9

1. Joseph Klatzmann, *Nourrir dix milliards d'hommes*, PUF, 1983, pp.39-57.
2. Jacques Giri, *L'Afrique en panne*, Karthala, 1986, p.42.
3. Roy Prosterman, *Forum du développement* No.117, April 1986, p.15.

Page 11

1. Herbert de Souza, quoted by Marcel Marloie, in *Courrier de la Planète* No.16, May 1993.
2. Yves Lacoste, *Contre les anti-tiers-mondistes*, La Découverte, 1985. p.45.
3. Frédérique Goubault, *La Lettre de Solagral* No.21, December 1983, p.4.
4. Cepa, 'La Faim en Argentine', *La Lettre de Solagral* No.47, April 1986, p.3.
5. *Scientific American*, quoted by Jean-Pierre Berlan in 'La logique infernale des rendements agricoles', *Le Monde*, 14 June 1988.

Page 13

1. 'Céréales et produits céréaliers en Méditerranée', in *Options méditerranéennes*, Etudes series, IAMM, December 1986.
2. Mahmoud Allaya et al., *Alimentation et agriculture en Méditerranée*, CIHEAM-Publisud, 1984, pp.54-70.
3. Philippe Chalmin, 'L'Egypte peut limiter sa dépendance', in *CCFD Magazine* No.55, January 1989, pp.6-7.

Page 15

1. Jacques Giri, 'Le Sahel est-il condamné à la famine?' in *Asie-Afrique: Greniers vides, greniers pleins*, Economica, 1986, p.130.
2. René Dumont, 'Agriculture', in *L'état du tiers-monde*, La Découverte, 1989, p.162.
3. Jacques Giri, *Le Sahel au XIXe siècle*, Karthala, 1989, pp.86-90.

Page 17

1. Vincent Leclercq, *Conditions et limites de l'insertion du Brésil dans le marché international du soja*, INRA-ESR Montpellier, 1987, p.310.

2. Paul Bairoch, *Commerce extérieur et développement économique de l'Europe au XIXe siècle*, Mouton, 1976, p.299.

Page 19

1. *L'Europe verte*, No.213, July 1985, pp.7-8.

Page 21

1. David Rosenberg, 'Expanded EC could hurt Israel', in *Jerusalem Post*, 28 April 1987, p.8.

Page 23

1. Yves Gazzo, *La Politique agricole de Etats-Unis*, Economica, 1984, pp.6-7.
2. Giovanni Anania et al., 'La bataille du blé', in *Courrier de la Planète* No.1, November 1991, p.29.
3. Alain Labrousse, Conférence Bioforce-Bordeaux, 1988.

Page 25

1. P. Bye and A. Mounier, 'Autonomie ou dépendance alimentaire: le cas de la Corée du Sud et du Japon', in *Economie et Humanisme*, No.270, March-April 1983, p.31-42.

Page 27

1. Thierry de Montbrial, *Ramsès 86-87*, Atlas Economica, 1986, pp.147-158.
2. Michel Bourdon, *Economie Rurale*, No.212, November-December 1992.

Page 29

1. Sylvie Bonny, *L'Energie et sa crise de 1974 à 1984 dans l'agriculture française*, INRA-ESR Grignon, 1986, p.497.
2. INSEE, 'Structure de la consommation en l'an 2000', in *Economies et finances agricoles*, February 1987, p.7.
3. Jean-Paul Charvet, *La Guerre du blé*, Economica, 1988, p.200.

Page 31

1. Philippe Chalmin, *Economie et finances agricoles* supplement, June-July 1983, p.31.
2. CFCE, *Les firmes alimentaires étrangères*, CFCE/DPA/IFE, 1987, p.109.
3. Salma Oncuoglu in *Courrier de la Planète*, No.6, April 1992, p.16.
4. Michel de Saint-Albin, 'L'origine et les effets de la concentration des industries alimentaires américaines', in *Agra Alimentation* No.1007, 19 December 1985.

Page 33

1. All data on this page extracted from *Agrodata*, Agropolis-Stratégies, Montpellier, 1991.

Page 35

1. Chantal Ducos, 'Du paysan aux multinationales', in *La Lettre de Solagral* No.36, April 1985, pp.7-10.
2. Chantal Ducos and Pierre Benoit-Joly, *Les Biotechnologies*, La Découverte, 1988, p.128.
3. The following section is extracted from *Biodiversity: What's at Stake?*, Catholic Institute for International Relations (CIIR Comment series), September 1993, p.16.

   The text of the Biodiversity Convention makes repeated use of conditional phrases such as 'on mutually agreed terms' even though the issues at stake are global and two parties alone cannot reach sustainable solutions. Bilateral deals pit rich Northern governments or corporations against poor Southern governments and farmers. A more equitable South-North balance requires broad multi-country negotiation.

   Stimulated by the bilateral approach of the Convention, some unsuspecting governments are considering offering their biological heritage to the highest bidder. This may hold the incentive of quick and easy income, but in the long term it points to the massive and unprofitable selling out of biodiversity. Southern governments would have much stronger negotiating power with the biotechnology laboratories of the North if they acted collectively. Yet companies are striking deals with individual countries across Africa, Asia

and Latin America. In this way local people are unlikely to be consulted on their governments' negotiations, unlikely to share in any monetary or scientific returns and unlikely to see any improvement in their livelihoods.
4. Ibid.
5. Albert Sasson, 'La conservation des ressources végétales', in *La Recherche* No.181, October 1986, pp.1,282-1,293.

Page 37
1. Philippe Chalmin, *Les Marchés mondiaux des matières premières*, PUF, Que sais-je?, 1984, pp.30-33.
2. Philippe Chalmin and Jean-Louis Gombeaud, *Cyclope – Les marchés mondiaux 86-87*, Economica, 1987, p.16.
3. Eric Sottas, L*es Pays les moins avancés*, Unctad INF-PUB 84-2, p.15.

Page 39
1. Gérard de Bernis and Jacques Léonard, 'Une ruineuse course aux crédits extérieurs', in *Le Monde Diplomatique* No.385, April 1986, pp.15-16.
2. Christian Domptin, 'La dette joue les prolongations', in *Alternatives Economiques* No.94, February 1992, pp.34-36.
3. Agir Ici, *La COFACE, un système d'assurance d'exportation à réformer*, September 1991, p.24.
4. S. Erbe, quoted by Jean Coussy and Philippe Hugon, 'Trois pays, trois types de contraintes', in *Le Monde Diplomatique* No.385, April 1986, p.17.
5. Philippe Norel and Eric Saint-Alary, op. cit., p.33.

Page 41
1. 1985 Facts, quoted by Bertrand Le Gendre, in *Le Monde*, 7 June 1988, p.49.
2. 'Le développement et la Suisse', in *Vers un développement solidaire* No.89, April 1987, p.17.

Page 43
1. Maria E. Cosio-Zavala, 'Mexique, Mexico, Mexicains', in *Populations et Sociétés* No.236, June 1989, p.3; Pierre Coulomb, 'L'exploitation familale en question', in *Les Paradoxes de l'agriculture française*, L'Harmattan, 1985.
2. Marcel Marloie, 'Paysans dans l'agro-alimentaire: nouvelles donnes', in *Agriculteurs en recherche* No.35, March-April 1987, pp.23-25.

Page 45
1. Jeanne Bisilliat and Clio Presvelou, F*emmes et politiques alimentaires*, ORSTOM, 1985, pp.699-712.
2. Annette Corrèze, 'Stratégies alimentaires, stratégies féminines, interventions de développement', ibid., p.393-404.
3. Marie Monimart, *Femmes du Sahel*, Karthala, 1989, p.263.
4. Jean Gallais, 'Milieu naturel et action de l'Etat au Sahel', in *Asie-Afrique: greniers vides, greniers pleins*, Economica, 1986, p.124.
5. Thierry Brun, 'Malnutrition et médiocre état de santé de la paysannerie sahélienne', in René Dumont, P*our L'Afrique j'accuse*, Plon, 1986, p.315-328.
6. Georges Courade, 'La sécurité alimentaire implicite: les femmes hors-développement au Cameroun', in *Femmes et politiques alimentaires*, ORSTOM, 1985, pp.371-392.

Page 47
1. Radha Sinha, *Le paysannat sans terre*, FAO, 1985, p.16.
2. Yves Lacoste *Contre les anti-tiers-mondistes*, la Découverte, 1985, p.135.
3. Bernard Kervyn, *Agir avec les sans-terre du Bangladesh*, Frères des Hommes, August 1983, p.16.

Page 49
1. Laurence Tubiana, 'Blé', in *Courrier de la Planète* No.1, November 1991, p.23.
2. R.Paquette, 'Rapidité des changements dans les habitudes de consommation alimentaire à

Kingston en Jamaique', quoted by Jean-Paul Charvet in *Les Greniers du Monde*, Economica, 1985, p.12.
3. 'Echographies', in *Réseaux* No.42, April-May 1987, p.3.

Page 51
1. Philippe Pons, 'Forêts: 125,000 kilomètres carrés disparaissent chaque année', in *Le Monde*, 18 June 1985, pp.21-23.
2. Daniel Dory, 'Comment enrayer la disparition des forêts tropicales', in *Le Monde Diplomatique* No.395, February 1987, pp.30-31.
3. Gérard Buttoud and Amani J.Kouame, 'L'évolution du commerce international des bois africains', in *Afrique contemporaine* No.134, April-June 1985, pp.20.
4. World Resources Institute, quoted in *Cyclope 88-89*, Economica, 1989, p.194.
5. ACP-EEC Assembly, quoted by J.Adnet et al., *Développement et environnement dans la politique de la Communauté européenne*, Creare, 1986.
6. Bernard Duhamel, 'Crise énergétique et crise environnementale en Afrique', in *Environnement africain* No.20-22, 1985, pp.85-106.
7. Jean-Roger Mercier, *L'Adieu aux arbres*, Syros, 1986, p.95.
8. Susan George, *L'Effet Boomerang*, La Découverte, 1992, p.40.

Page 53
1. US Geographical Survey Circular 1004, 1988.
2. Yves Gazzo, *La politique agricole des Etats-Unis*, Economica, 1984, p.90.
3. *Pollution des eaux par les engrais et pesticides*, OECD, 1986, pp.33-45.
4. Roger Cans, 'Eau non potable', in *Le Monde*, 22 January 1986.
5. 'L'eau, les nitrates et la santé', special issue of *Eaux et Rivières de Bretagne* No.59, 1987, p.30.

Page 57
1. Alain Mounier, 'Le "péril blanc". Les agricultures des pays riches: une menace pour le tiers-monde', in *Agricultures en question* No.6, 1984, pp.73-93.

Page 59
1. Marcel Marloie, 'Les leçons du XIXe siècle et les dangers du dogmatisme', in *Le Monde Diplomatique* No.382, January 1986, p.20.
2. Marcel Marloie, 'Tourteaux: de l'héritage colonial à la relance des concurrences', in *La Lettre de Solagral* No.26, May 1984, p.8.
3. Marcel Marloie, *L'internationalisation de l'agriculture française*, Ed. ouvrières, 1984.

Page 61
1. Marcel Marloie, 'Les leçons du XIXe siècle et les dangers du dogmatisme', in *Le Monde Diplomatique* No.382, January 1986, pp.20-21.
2. Denis Sautier, 'Pourquoi la Thailande?', in *La Lettre de Solagral* No.2, November 1981, p.11.

Page 65
1. Philippe Evrard, in *Les Pays industrialisés souhaitent-ils ou non éliminer leurs agriculteurs?* CEF-PIA, 1987, pp.5-14.
2. Commissariat général du Plan, *Pour un role actif de la CEE sur la scène agricole internationale*, La Documentation française, 1985, pp.157-63.

Page 67
1. Philippe Chalmin, 'Le rêve et la réalité', in *Alternatives Economiques* No.47, May 1987, pp.15-18.
2. Antoine Labey, 'Accords perdus', in *Courrier de la Planète* No.16, May 1993.
3. Anne-Sophie Boisgallais and Charles Condamines, *Le sucre: le Nord contre le Sud*, Syros, 1988, p.151.

Page 69
1. Jean-Paul Charvet, *Les Greniers du monde*, Economica, 1985, pp.39-40.

2. *Agra Alimentation* No.1065, 26 March 1987, pp.8-9.
3. Philippe Chalmin and Jean-Louis Gombeaud, *Cyclope – Les marchés mondiaux 88-89*, Economica, 1989, p.161.
4. 'Pepsi International Urged to Assist Sugar Workers', in *Sugar World* Vol.10, No.12, April 1987, pp.5-6.

Page 71
1. With the notable exception of the Feruzzi group. See *Trans Rural Express* No.113, 22 June 1989, p.4.
2. Dan Morgan, *The Merchants of Grain*, Penguin, 1980.
3. Vladimir Andreff, *Les Multinationales*, La Découverte, 1987, p.125.

Page 73
1. Edgard Pisani, *La Main et l'Outil*, Robert Laffont, 1984, pp.89-90.
2. Including transport costs and retail profit margins. See *World Development Report*, World Bank, 1986.
3. Jacques Giri, *L'Afrique en panne*, Karthala, 1986, pp.60-70.

Page 75
1. Jacques Giri, op. cit., pp.76-78.
2. The IMF said the measure would boost export promotion, investment and economic competitiveness. However, the sudden nature of the move – it came after 46 years without any change to the fixed rate – led many observers to predict that several West African countries would face acute difficulties in adjusting. As well as strong inflationary pressure, a weaker currency would make food imports and domestically produced basic commodities more expensive for consumers, thus threatening the possibility of social unrest.

Page 77
1. Vincent Leclercq, 'Le Brésil met le FMI sur la touche', in *La Lettre de Solagral* No.48, May 1986, pp.7-11.
2. Jean Coussy, 'Les méandres de la rigueur', in *La Lettre de Solagral* No.82, June 1989, pp.4-5.
3. Michel Camdessus, in *La Vie* No.2349, 28 May 1992, p.31.

Page 79
1. Paul Bairoch, *De Jéricho à Mexico*, Gallimard, 1985, pp. 636-60.
2. *Rapport sur l'alimentation mondiale*, FAO, 1985, p.11.
3. Denis Sautier, 'Est-on mieux nourri à la campagne ou dans les bidonvilles?', in *La Lettre de Solagral* No.56, February 1987, p.11.
4. Thierry Brun, 'Malnutrition et médiocre état de santé de la paysannerie sahélienne', in René Dumont, *Pour l'Afrique j'accuse*, Plon, 1986, p.319.

Page 81
1. Albert Sasson, *Nourrir demain les hommes*, Unesco, 1987, pp.419-20.
2. A. Basler, 'Exportations agricoles et déficit alimentaire en Afrique', in *Economie Rurale* No.173, May-June 1986, pp.60-66.
3. Claude Freud, *Quelle coopération?*, Karthala, 1988, p.103.
4. Jacques Chonchol, *Le Défi alimentaire*, Hachette, 1987, p.45.

Page 83
1. Solagral, *L'Aide alimentaire*, Syros, 1984, p.135.
2. Y. Hossein Farzin, 'Food Import Dependence in Somalia', *World Bank Discussion Paper* No.23, 1988, p.35.
3. Stéphane Jost, 'Aide alimentaire, blé ou mil', in *La Lettre de Solagral* No.57, March 1987, p.13.

Page 85
1. Jean-Claude Romain, 'Marché céréalier: une affaire de grands', in *La Lettre de Solagral*

No.30, October 1984, pp.7-9.
2. *Poverty and Hunger*, World Bank, 1986.

Page 87
1. François Bellec, 'Les pêcheurs philippins reprennent espoir', in *Pêcheurs des tiers-mondes*, CCFD-Solagral, 1986, pp.54-59.
2. J. Weber and A. Fontana, 'Pêches et stratégies de développement, discours et pratiques', in *Le Courrier ACP-CEE* No.85, May-June 1984, pp.74-77.
3. Alain Le Sann, 'Mer d'avenir', in *Pêche et developpement* No.20, 2nd quarter 1992, p.1.

Page 89
1. Chantal Ducos and Pierre Benoit-Joly, *Les Biotechnologies*, La Découverte, 1988, p. 127.
2. Michel Pimbert, 'Les Biotechnologies appliquées à l'agriculture', in *La Lettre de Solagral* No.36, April 1985, p.14.
3. Jacques Grall, 'Les moyens du miracle', in *Croissance des jeunes nations* No.287, October 1986, p.50.
4. Albert Sasson, 'La conservation des ressources végétales', in *La Recherche* No.181, October 1986, p.1289.

Page 91
1. Frederic Clairmont and John Cavanagh, *Alcool et pouvoir des transnationales*, P.M. Favre-CETIM, 1986, p. 247.
2. *Unomásuno*, Mexican daily newspaper quoted in *La Lettre de Solagral* No.10, October 1982, p.23.
3. WHO, 'La consommation de cigarettes dépasse la croissance démographique', in *Santé du Monde*, March 1987, p.30. *Development Forum* No.137, September-October 1988.

Page 93
1. 'L'environnment contre les pauvres?', in *La Lettre de Solagral* No.58, April 1987, pp.8-11.
2. Commission indépendante sur les questions humanitaires internationales, *Famine: mieux comprendre, mieux aider*, Berger-Levrault, 1986, pp.84-95.
3. Phil O'Keefe, in *State of India's Environment*, CSE, Delhi, 1985, p.180.
4. Marc Dufumier, communication to the 'Agronomes sans frontières' talk, Paris, 13 June 1987.
5. Lloyd Timberlake, *Africa in Crisis*, Earthscan, 1985.
6. Christian Kuchli, *Des forêts pour les hommes*, Payot, 1984, pp.17-47.

Page 95
1. Marc Dufumier, *Les politiques agraires*, PUF, Que sais-je? 1986, pp.87-120.
2. Claude Aubert, 'Les réformes agraires ou la genèse incertaine d'une nouvelle voie chinoise', in *Revue Tiers-Monde* No.108, October-December 1986, pp.727-754.
3. Tran Ngoc Bich, 'Vietnam: une économie à la dérive', *Faim Développement Dossiers* No.87-84, April 1987, p.11.
4. Bertrand Delpeuch, *Pour des agricultures paysannes*, Fondation pour le Progrès de l'Homme, March 1989, p.105.
5. Miran Dolot, *Les Affamés, l'holocauste masqué*, Ramsay, 1986, p.333.

Page 99
1. Edgard Pisani, 'Au-delà des impératifs purement gestionnaires', in *Le Monde Diplomatique* No.382, January 1986, p.20.
2. François de Ravignan, *La Faim, pourquoi?* Syros, 1983, p.125.
3. Serge Michailof, 'Il est temps de réorienter notre action', in *Courrier de la Planète* No.16, May 1993.
4. René Dumont, *Taiwan, le prix de la réussite*, La Découverte, 1987, p.174.

Page 101
1. Marc Dufumier, 'La question agraire en Amérique centrale', in *Nicaragua aujourd'hui*

No.46-47, summer 1988, pp.12-16.
2. Jacques Chonchol, *Le Défi alimentaire*, op. cit., p.252.
3. François de Ravignan, 'L'espoir déçu des paysans andalous', in *Le Monde Diplomatique* No.410, May 1988, p.8.

Page 103
1. André Pochon, *La Prairie naturelle à base de trèfle blanc*, ITEB, 1981.
2. Jean-Luc Messe, 'L'agriculture biologique, une solution pour le tiers-monde?', in *Nature et Progrès*, 1987, p.15-18.

Page 105
1. Bertrand Delpeuch, 'Porcs rustiques et choix de développement', in *L'Etat du Tiers-Monde*, La Découverte, 1989, p.161.
2. Inaremac, 'Sélectionner le maïs local', *La Lettre de Solagral* No.6, 2nd quarter 1982, p.23.
3. M. Molenat and C. Legault, 'Le porc dans les pays en voie de développement', in *B.T.I.* No.406, 1986, pp.39-53.
4. André Briend, *Prévention et traitement de la malnutrition*, ORSTOM, 1985, pp.21-24.

Page 107
1. Alain Le Sann et al., 'L'état de l'environnement en Inde', in *La Lettre de Solagral* No.58, April 1987, p.12.
2. Jeanne Bisilliat et al., *Femmes et politiques alimentaires*, ORSTOM, 1985, p.741.
3. Vina Mazumdar, 'Un autre développement avec les femmes: point de vue d'Asie', *Development Dialogue*, 1985, p.90.

Page 109
1. Annie Simon, 'Les aides triangulaires', in *Actuel Développement* No.70, January-February 1986, p.5.
2. Stéphane Jost, 'Repeindre l'Afrique en vert', in *Alternatives Economiques* No.48, June 1987, p.23.

Page 111
1. Christian Domptin, 'Les leçons de la crise', in *Alternatives Economiques* No.94, February 1993.
2. Anne-Sophie Boisgallais, 'L'aide tentée par le troc', in *Alternatives Economiques* No.94, February 1992, p.37.
3. Christine Poupon, 'Ethique ou marketing', in *Courrier de la Planète* No.16, May 1993.

Page 113
1. Anne-Marie Bienvenu et al., 'Alimentation des nourrissons: un code international', in *La Lettre de Solagral* No.1, October 1981, pp.7-16.
2. Jean-Claude Buffle, *N... comme Nestlé*, Alain Moreau, 1986, p.460.

Page 115
1. Alain Le Sann et al., *La Pirogue et le chalutier*, CCFD-Solagral, 1986, p.117.
2. Radha Sinha, *Le paysannat sans terre*, FAO, 1985, p.79.
3. Edgard Pisani, La Main et l'Outil, Robert Laffont, 1984, p.100.

Page 117
1. Bertrand Delpeuch, 'Politisons nos désirs de solidarité', in *La Lettre de Solagral* No.50, July-August 1986, p.17.

Farmers' Link

Farmers' Link is an independent, non-profit making association with the following aims:

- to promote awareness of agricultural problems faced by farmers worldwide;
- to deepen understanding of the links between farmers and landowners in the 'North' and the 'South' of the world;
- to widen the debate on the social and environmental implications of food production;
- to explore possible changes in agricultural policies which would reflect the common interests of farmers and rural communities worldwide;
- to engage in activities which promote a transition towards sustainable agriculture and rural development, both in the UK and the rest of the world.

Farmers' Link plays a key role in the UK in the debate over the future of farming and rural development. In 1994 we started a programme in the East Anglian farming community: 'Education for Sustainable Agriculture and Rural Development'. Based on the assumption that we have much to learn from the so-called 'Third World', our partners in this programme include the Windward Islands Farmers' Association, the Coordination of Latin American Banana Workers' Unions, the Zimbabwe Farmers' Union and the Democratic Peasant Farmers' Movement of the Philippines.

By acting locally, but also regionally and nationally, we hope to play our part in a global transition to 'sustainable development'.

Seed and Surplus: An illustrated guide to the world food system

If you teach young people or adults about world food issues and would like to use the graphics which appear in this book, you will be interested to know that a set of 56 transparencies for use with an overhead projector is available from **Farmers' Link, 38-40 Exchange Street, Norwich NR2 1AX.**
Tel 0603 765670, Fax 0603 761645, E-mail = gn:nead.

CIIR

Catholic Institute for International Relations

CIIR works to overcome poverty and injustice in the Third World. Founded in 1940, it is an independent charity which works with people of any religious belief or none.

Tackling the issues

CIIR's **Education Programme** provides information on socio-economic, political, human rights and church issues in the Third World. It has sections working on Asia, Latin America and Southern Africa, and one which analyses the international causes of Third World poverty.

Sharing Skills

International Cooperation for Development, CIIR's overseas programme, recruits a wide range of experienced professionals to share their skills in small-scale development projects in Latin America, the Caribbean, Southern Africa and the Middle East.

If you would like to know more about CIIR, or if you would like to join CIIR, please contact the **Membership Secretary, CIIR, Unit 3, Canonbury Yard, 190a New North Road, London N1 7BJ, UK.**
Tel 071 354 0883. Fax 071 359 0017.
Membership: UK £15, unwaged £5, Overseas £20.

CIIR members receive *CIIR News* four times a year, CIIR's *Annual Review*, and a range of pamphlets, including titles in the *Comment* series. Members are offered special discounts on CIIR publications.

147591